New Functional Materials
and Emerging Device Architectures
for Nonvolatile Memories

MATERIALS RESEARCH SOCIETY
SYMPOSIUM PROCEEDINGS VOLUME 1337

New Functional Materials and Emerging Device Architectures for Nonvolatile Memories

Symposium held April 25–29, 2011, San Francisco, California, U.S.A.

EDITORS

Dirk J. Wouters
IMEC
Leuven, Belgium

Eisuke Tokumitsu
Tokyo Institute of Technology
Yokohama, Japan

Orlando Auciello
Argonne National Laboratory
Argonne, Illinois, U.S.A.

Panagiotis Dimitrakis
NCSR Demokritos
Aghia Paraskevi, Greece

Yoshihisa Fujisaki
Hitachi Ltd.
Tokyo, Japan

Materials Research Society
Warrendale, Pennsylvania

CAMBRIDGE UNIVERSITY PRESS
Cambridge, New York, Melbourne, Madrid, Cape Town,
Singapore, São Paulo, Delhi, Tokyo, Mexico City

Cambridge University Press
32 Avenue of the Americas, New York, NY 10013-2473, USA

www.cambridge.org
Information on this title: www.cambridge.org/9781605113142

Materials Research Society
506 Keystone Drive, Warrendale, PA 15086, USA
http://www.mrs.org

First published 2011

CODEN: MRSPDH

ISBN: 978-1-60511-314-2 Hardback

CONTENTS

ADVANCED FLASH AND NANO-FLOATING GATE MEMORIES

*Invited Paper

RESISTIVE SWITCHING MEMORIES

*Invited Paper

PREFACE

Symposium Q, "New Functional Materials and Emerging Device Architectures for Nonvolatile Memories" held April 25–29 at the 2011 MRS Spring Meeting in San Francisco, California, was a follow up of a previous series of related Symposia on Non-Volatile Memories at MRS Annual Meetings, including the following: 1) Symposium "**Materials and Physics for Non-Volatile Memories**" organized for the first time in 2004 during the Fall MRS Meeting in Boston; 2) Symposium "**Materials and Processes for Non-Volatile Memories**" held in the 2007 MRS Spring Meeting; 3) the Symposium series titled "**Materials Science and Technology of Non-Volatile Memories**" held at the 2006 and 2008 MRS Spring Meetings; 4) the Symposia entitled "**Materials and Physics for Non-Volatile Memories**" held in the 2009 and 2010 MRS Spring Meetings.

The high attendance and large paper submission (in total 67 oral and 34 poster contributions were presented in 10 sessions, in addition to 9 invited talks), indicate the continuing strong international interest and research effort in the field of emerging new non-volatile memory materials. The Session on Phase Change Memories was shared with the Symposium Phase-Change Materials for Memory and Reconfigurable Electronics Applications. We are also proud that two of the presented posters of this Symposium have won the *MRS Poster Award*.

Main areas of research featured in Symposium Q were Advanced Flash Memories, Phase Change Memories, and–in particular–Resistive Switching Memories. In addition, Ferroelectric Memories, Organic Memories, and New Emerging Memories remained of interest.

The selected papers in this Proceedings volume have been categorized in three Chapters. The Chapter *Advanced Flash and Nano-Floating Gate Memories* deals with solutions for scaled Flash memory, including the use of new high-κ layers and nanocrystals. *Resistive Switching Memories* are discussed in the second Chapter. The final Chapter includes Phase Change and Ferroelectric Memories, as well as contributions on Organic Memories.

A highly successful one-day tutorial was conducted, including tutorials on Memory Devices for Organic Electronics, Ferroelectric Memories, Redox-based Resistive Switching, and Materials and Concepts for Magnetic Memory.

With international contributions from university, research centers and industry, the papers from this Symposium Proceedings reflect the recent advances in material science and their influence in the memory technologies addressed in this Symposium.

Dirk J. Wouters
Eisuke Tokumitsu
Orlando Auciello
Panagiotis Dimitrakis
Yoshihisa Fujisaki

August 2011

ACKNOWLEDGMENTS

The symposium organizers would like to thank the tutorial speakers as well as the invited speakers who contributed to the success of this Symposium:

Tutorial Speakers:
- Michael C. Petty (Durham University, United Kingdom), Daisaburo Takahsima (Toshiba Corporation, Japan), Rainer Waser (RWTH Aachen University, Germany), Samuel D. Bader (Argonne National Laboratory)

Invited Speakers:
- Kirk Prall (Micron Technology), Ted Moise (Texas Instruments), S. Sakai (National Institute of Advanced Industrial Science and Technology, Japan), Seungbum Hong (Argonne National Laboratory), Andrea Redaelli (Numonyx), Hongsik Jeon (Samsung Electronics), Ramamoorthy Ramesh (University of California, Berkeley), Tsu-Jae King Liu (University of California, Berkeley), Carlos Paz de Araujo (University of Colorado at Colorado Springs and Symetrix Corporation)

We also wish to thank the following organizations for their financial support of this Symposium:

- Annealsys
- Applied Materials Inc.
- M. Watanabe & Co., Ltd.
- Park Systems Corporation
- Universal Systems Co., Ltd.

MATERIALS RESEARCH SOCIETY SYMPOSIUM PROCEEDINGS

MATERIALS RESEARCH SOCIETY SYMPOSIUM PROCEEDINGS

Prior Materials Research Society Symposium Proceedings available by contacting Materials Research Society

Advanced Flash and Nano-Floating Gate Memories

Mater. Res. Soc. Symp. Proc. Vol. 1337 © 2011 Materials Research Society
DOI: 10.1557/opl.2011.1028

Scaling Challenges for NAND and Replacement Memory Technology

Kirk Prall
Micron Technology, 8000 S. Federal Way, Boise, ID 83716
kprall@micron.com

ABSTRACT

Planar NAND technology is rapidly approaching its fundamental limits and will likely transition to a three dimensional structure. The scaling challenges facing NAND will be reviewed. Emerging memory technologies, such as the cross-point, will be discussed. The materials challenges facing emerging memories will be reviewed.

INTRODUCTION

NAND is currently the dominant semiconductor non-volatile storage technology used in solid state discs, USB flash cards, camera cards, etc. NAND technology has been rapidly scaling, faithfully following Moore's law since its invention circa 1987 by Toshiba [25].

NAND is facing severe scaling challenges which will likely result in the end of planar silicon scaling at around the 15nm generation. NAND may survive by adopting a 3-D structure, continuing to increase density, and reducing cost by adding layers in the third dimension. A possible alternative path may be that NAND will be replaced by a technology operating on a new mechanism that departs from silicon based electron storage. Large materials challenges exist and major breakthroughs will be required to successfully commercialize a NAND successor.

NAND SCALING

Currently, NAND manufacturers are shipping high volumes of product in the 20-25nm range [1][26]. There is a large research effort to push planar NAND technology to its ultimate limit. The amount of money expended on NAND research and development yearly is in the range of $750M-$1B; giving the technology a huge amount of technical momentum and enabling NAND to outrun other potential technologies. The technical challenges facing NAND are daunting and will ultimately lead to the demise of planar NAND as we know it.

The technical challenges facing NAND scaling are summarized in Table I. There are several categories to consider: fundamental capacitance limits caused by scaling, scaling issues caused by the inability to reduce voltages or dielectric thickness, and reliability challenges.

Table I NAND Scaling Challenges

Capacitance Issues	Ref #
Floating gate capacitance trending to zero	[1][2]
Electrons per cell trending to zero	[1][2]
Interference (parasitic capacitance trending to 100%)	[3][4][5]
Scaling Issues	
Tunnel oxide thickness stuck at ~7nm	[6]
Interpoly dielectric thickness stuck at ~10nm	[7]
Cell operating voltages stuck at ~25V (MLC)	[1][2]
Isolation stuck at ~6-8V, high aspect ratio isolation	[1][2]
Inhibit stuck at ~10V	[8]
Wordline to Wordline field trending to >10 MV/cm	[1]
Variation increasing	[9][10][11]
Noise increasing	[12][13]
Quantum mechanical tunneling noise	[14]
Parasitic electron trapping	[1]
Reliability Challenges	
Program disturb	[15][16]
Trapping / detrapping	[17]
Quick electron detrapping	[18]
Stress induced related charge loss	[6]
Retention is degrading	[19],[20]
Read disturb	[21]
Cycling is degrading	[22]
Random telegraph signal noise	[23]
Increasing ECC requirements	[1][24]

Several of these scaling challenges will be discussed in the following sections in order to highlight the difficulty of the problems that NAND is facing. The topics covered will include the few electron problem, parasitic electron trapping, giant random telegraph signal (RTS) noise, and quantum mechanical tunneling variation.

Few Electron Problem

The number of electrons on the floating gate is given by the equation $Q=CV$, where C is the inter-poly capacitance and V is the floating gate voltage. The inter-poly dielectric has been relatively constant over many generations, consisting of an oxide/nitride/oxide sandwich. The dielectric thickness and composition is limited by charge loss through the dielectric causing retention failures, and charge leakage during programming limiting the maximum threshold voltage of the transistor. Attempts to use high dielectric constant materials to increase the inter-poly capacitance have not been successful to date. The capacitor area of the floating gate scales with the area of the memory cell decreasing by the square of the shrink in feature size. The third dimension of the floating gate is the height and is limited by the practical limits of the aspect ratio and toppling of the floating gate stack. In practice, the floating gate capacitor area shrinks with scaling and the number of electrons stored on the floating gate decreases with scaling, as shown in Fig. 1. In 25nm MLC NAND technology, 300-500 mV (30-50 electrons) of signal separate the states stored on the cell as a transistor threshold voltage, which increases the difficulty of maintaining state separation. The small number of electrons stored on the floating gate of highly scaled devices impacts many aspects of the cell, as many mechanisms are degraded by the lack of electrons.

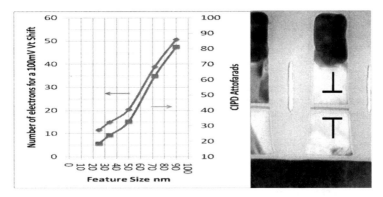

Fig. 1. The floating gate interpoly dielectric capacitance is shown as a function of the feature size highlighting the reduction in the capacitance with scaling (right axis). The left axis of the graph shows the number of electrons required to shift the threshold voltage of the cell by 100mV. The right picture shows a cell cross-section identifying the interpoly capacitance.

Parasitic Electron Trapping

The electric fields of a cell dielectric during operation are very high, typically reaching levels as high as 10 MV/cm. The high field gives the electrons high energy and can cause the electrons to be trapped at parasitic locations in the cell instead of the intended location in the floating gate, where the trapped electrons give a controlled shift in the floating gate potential in order to store the state of the cell. Fig. 2 shows the possible locations in the cell where the parasitic charge can be trapped. The table in Fig. 2 shows the number of electrons needed to shift the threshold of the floating gate voltage by 100mV. The impact of parasitic charge trapping can be obtained by comparing the number of electrons in the parasitic locations with the number of electrons on the floating gate (Q_{fg} in green). The numbers shown in red require a smaller number of electrons to shift the cell threshold voltage than the charge placed on the floating gate. Therefore the red locations can dominate and degrade the cell operation. Note that the number of dominant trapping locations increases with scaling. For the cell to operate properly, the trapping ability of the parasitic locations has to be minimized by careful optimization of the tunnel, interpoly, sidewall oxides, and junctions.

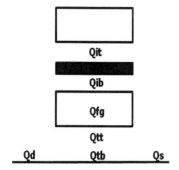

	50nm	35nm	25nm
Qtb	4	2	1
Qtt	9	7	4
Qib	22	17	9
Qit	149	103	100
Qs	33	9	5
Qd	61	16	10
Qfg	18	12	10

Fig. 2. The left sketch shows potential parasitic electron trapping locations in a NAND cell. The right table shows the number of electrons required at each parasitic location for a 100mV shift in threshold voltage based on TCAD electro-static simulations.

Giant Random Telegraph Signal Noise

As can be seen in the previous section, a smaller number of electrons can have a large impact on a scaled NAND cell. In fact, a single electron can shift the threshold voltage and cell current, resulting in significant noise that can disrupt the sensing and operation of the cell. RTS is typically caused by an electron trap modifying the channel conduction and mobility, due to the charge impact of trapped electrons on the movement of the channel electrons. However, in reality, the channel current can be observed as filamentary in small devices. The filamentary current flow is governed by several competing mechanisms. First, the dopant boron atoms in the channel reduce the mobility and the filamentary current tends to flow where the boron atoms are nonexistent. Secondly, electron traps and interface states at the surface degrade the mobility and the filamentary current flow tends to avoid the filled traps and interface states. In a scaled device the channel electron flow forms a filament as the charge flow avoids these defects and a majority of the current flow can be through the filament. When a trap above the conducting filament fills or empties, a large change in the current occurs and creates the giant RTS effect causing a disproportionately large impact on the channel current. Giant RTS creates time varying current noise in the cell current, which causes very detrimental effects on the cell operation. The large variation in the threshold voltage is shown in Fig. 3, which is typically recovered by error correction used in NAND operation.

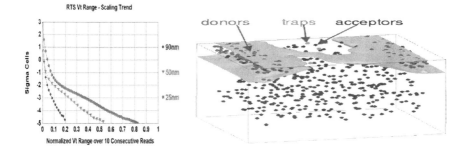

Fig. 3. The graph shows the variation in Vt of a distribution of cells measured 10 times to show the time varying giant RTS effect. Note the degradation in RTS with scaling. The image on the right shows a filamentary conducting channel in a scaled device. Asenov [10][11].

Quantum Mechanical Tunneling Variation

NAND cells program and erase via quantum mechanical (Fowler-Nordheim) tunneling across the tunnel oxide. The few electron problem requires careful control of the number of electrons that tunnel during each programming voltage pulse. For a scaled cell, the variation in the number of electrons injected with each programming pulse is observed as noise, which widens the threshold voltage distribution of the cells resulting in a degradation of usable threshold voltage (signal) in the cell. Essentially, each time a cell is programmed it behaves differently, making it difficult to hit a targeted threshold voltage.

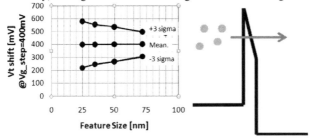

Fig. 4 . The degradation in cell function due to program noise is shown in the plot as an increase in the +/- 3σ distribution with scaling. The y-axis shows the threshold voltage shift of a distribution of cells with a constant 400mV programming pulse step. The programmed distribution of the cell increases with scaling due to variation in the Fowler-Nordheim tunneling, caused by the few electron problem. The image on the right shows a representation of electrons tunneling through the tunnel oxide.

7

POST NAND TECHNOLOGY

NAND technology is mainly driven by the lowest cost per bit. In order to achieve the lowest cost per bit, NAND technology sacrifices in many areas, such as: page addressing, large block sizes, complex controller technology requirements, poor latency, etc. Any NAND replacement technology will require a lower cost per bit, which is difficult to achieve given the current, highly-optimized position of state of the art NAND. The ability of NAND technology to store multiple bits per cell creates a large cost benefit that must be matched by any potential successor. There are potential paths where a future NAND replacement could have a different feature set, which is more valuable and could potentially reduce the cost pressure. Fig. 5 shows the likely future scaling paths for NAND [28].

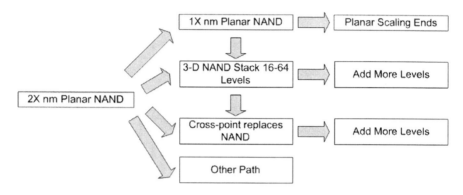

Fig. 5. Future scaling paths for NAND and potential replacement technologies.

As planar NAND reaches its limit, at around 15nm, its long run following Moore's law will come to an end, unless major breakthroughs in the problems shown in Table I are achieved. The most popular potential successor to planar NAND is 3-D NAND, where classical scaling will be replaced by stacking cells in the vertical direction in order to continue increasing density and reducing cost [27][29]. Instead of reducing the feature size, additional layers are stacked for cost reduction. Dozens of potential cell structures have been proposed. This is a logical path for NAND since the cell function and operation are very similar to planar NAND.

Cross-Point Memory
An alternative architecture to 3-D NAND is cross-point memory, which is also a 3-D structure. This type of memory is being heavily researched by many organizations. A sketch of cross-point memory is shown in Fig. 6.

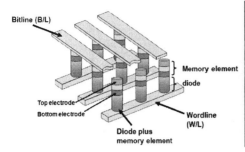

Fig. 6. Cross-section of cross-point memory [34].

Cross-point memory consists of two electronic elements and a metal interconnect (bitlines and wordlines). The electronic elements are a diode and resistive memory element (RRAM). The technical requirements of each of these elements are very stringent and difficult to achieve. There has been no commercially successful, rewriteable, cross-point memory as of this date.

RRAM Element

The RRAM memory element typically functions by a resistance change due to an applied voltage or current, which is used to define the logic states of the memory. Many different physical mechanisms are being investigated as potential RRAM elements. The more popular ones are shown in Table II.

The CMOS silicon based technology, which has been the mainstream technology for non-volatile memory for decades, is very well understood and can be easily modeled from first principles. Floating gate operation is very simple and is governed by the addition or removal of electrons from the floating gate. The replacement technologies shown in Table II operate on far more complex physical mechanisms, which tend to be poorly understood. Magnetic and phase-change based technologies have been researched for many years and are reasonably well understood. Most of the other resistive change based memories are poorly understood and it is not unusual to see multiple mechanisms described for identical materials systems in the literature. The situation is further complicated by differences in processing between different researchers.

Table II Mechanisms for RRAM Memory Element

RRAM Element	Mechanism	Ref #
Barrier Modulation	Modification of Schottky barrier interface	[31]
Conductive Bridge RAM	Metal filament formation through a dielectric	[31]
Ferroelectric tunneling	Polarization change modifies barrier height	[30]
Magnetic	Spin torque	[33]
Metal Oxide	Stochiometric change results in resistance change	[41]
Multi-valence Oxide	Ion motion changes resistance	[32]
Phase Change	Amorphous to crystalline transition	[34]

Case Study – Phase Change Materials

Some of the phase change mechanisms and properties will be reviewed in order to highlight the complexities of the new memory systems and to demonstrate the great deal of work required to reduce the technology to practice and to understand the material system well enough to make it commercially viable. Phase change materials have been heavily researched because of their applications in CDs/DVDs and, recently, in memory systems.

Phase change materials are a good case study for understanding the complexity of emerging memory material systems because it is the most mature material system. More than a decade of intensive research has gone into enabling phase change as a memory technology [36]. As a result, the important material properties that require optimization are understood. Phase change is the only emerging memory system currently in low volume production. Table III shows a partial list of the specific material parameters that must be optimized for successful phase change memory. The 23+ material properties and parameters must be optimized simultaneously to meet the demanding specifications for a memory cell. Clearly this is a daunting task.

Table III Important Phase-Change Material Properties for Memory Applications

Property	Description	Page Number [36]
Dopants and interface	Enhances nucleation probability stabilizes amorphous phase crystallization speed and retention crystallization temperature	88, 94, 326
Eutectic point	Controls nucleation or crystal growth	88
Glass transition temperature (related metric over large time scale)	Amorphous state stability crystallization temperature nucleation rate	94, 150
Hole concentration and mobility	Resistance	175
Incubation time	Makes amorphization possible	144
Large atomic displacement	Crystallization speed	215
Melting temperature (related long time scale metric)	Maximum crystalline rate between Tg and Tm	150
Oxidation	Interface failure	394
Reset energy	Power consumption cycling	378
Stoichiometry drift with cycling	Cycling failure - stuck crystal or glass state	206, 377
Structural relaxation and trapping	Drift reliability problem	320
Temperature dependence of resistivity	Maximum operating temperature	186
Tg/Tm glass formation	Good PCM materials are poor glass formers	151
Thermal conductivity of all cell materials	Cell operation efficiency set and reset times	118
Thermal vibrations	Required at high temp for fast operation	215-217
Vacancy percentage Cavities and voids	Ease of transition between states	27
Viscosity	Nucleation rate	131
Volume change between amorphous and crystalline states	Mechanical failure at electrode	207, 377

The material parameters can be placed in different categories related to their impact on the cell. Some of the parameters are fundamental to the cell operation. For example, chalcogenides have the very interesting property that the amorphous and crystalline states have nearly identical structure, which enables repeatable switching between the crystalline and amorphous states with reasonable energy expenditure. The chalcogenide structure has a large amount of vacancies and voids, which enables easy switching between the states. Other parameters control the cell performance. For example, the set speed is controlled by the nucleation and growth rates, which set limits on the speed of data movement in the memory. Some of the parameters control the reliability of the cell. For example, the volume change between the states can cause the electrodes to delaminate, inducing cell failure. Another example is any electrical or structure change causing a drift in resistance that limits the MLC capability of the cell. The other materials shown in Table III are as equally complex as the phase change system.

Diode Element

The second active component in a cross-point memory is the diode. The cross-point architecture places some unique requirements on the diode in order to enable cell function in a cross-point array. The off state leakage must be very low to prevent sneak-paths, and the current carrying requirements for some memory systems can be very high. The cross-point diode also has to be formed at low temperatures to be compatible with back end of line copper processing. The diode may be required to be symmetric if the RRAM element requires a positive and negative bias during operation. This makes the diode requirements more difficult. Most of the lower current, and more interesting RRAM cells, are bipolar and require a more complex diode. Table IV shows an example specification for a cross-point diode.

The diode in a cross-point structure is not necessarily a diode in the classic sense. It is sometimes referred to as a non-ohmic device (NOD). The search for a device capable of meeting the requirements for the cross-point diode has lead to a large search for different materials structures (typically layers of films) that can yield an I-V curve, which is beneficial for a cross-point structure. There is currently no published device that can meet the extreme requirements of a cross-point structure. Fig. 7 shows some of the more promising devices that have been reported. There is a large spectrum of devices that are being researched, each requiring a significant advancement in materials properties and capabilities. The diode requirements are closely related to the RRAM cell element and must be specifically designed to work with a given RRAM element.

Table IV Example Cross-point Diode (NOD) Requirements

Parameter	Specification
Ion	10^7 - 10^8 amps/cm^2 for high current cells 10^5 - 10^7 amps/cm^2 for low current cells
Ion/Ioff Ratio	$>10^4$
Von	1-4 volts
Switching speed	nS-uS
Slope	>1 decade/ 100mV
symmetric or asymmetric	depends on RRAM
Reliability	10^4-10^6 program erase cycles $>10^6$ read cycles
Temperature dependence	small - circuit must be able to compensate for temperature variation
Fabrication temperature	<400C to be compatible with BEOL processing temperatures

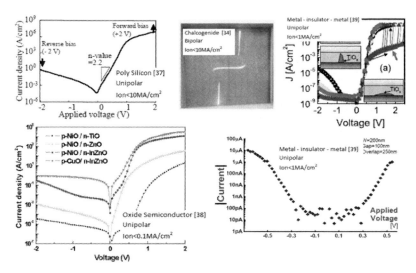

Fig. 7. I-V curves of a few of the more promising devices for cross-point diodes or non-ohmic devices (NOD) [42].

CONCLUSIONS

NAND technology is currently under strong pressure, due to scaling challenges. A large amount of research is under way to convert NAND to a three dimensional structure. Alternative memory architectures, such as the cross-point, are also being heavily researched. Large advancements in materials technology are required in order to enable a viable successor to NAND technology. The material science community will be challenged to produce the breakthroughs that will spawn a new generation of memory technology.

ACKNOWLEDGMENTS

The author gratefully acknowledges the Micron Emerging Memory Team who contributed greatly to the technical content and quality of this paper.

REFERENCES

[1] Prall, K. Parat, IEEE IEDM 2010, pg. 102
[2] Prall, NVSMW, 2007, pg. 5
[3] Lee, et. al., IEEE EDL, May 2002, pg 264
[4] Hung, et. al., IEEE TRED, April 2008, pg. 1020
[5] Ghetti, et. al., SSE, 2005, pg. 1806
[6] Okuyama, et. al., IEEE IEDM 1998, pg. 905
[7] Ho, et. al, IEEE EDL, Nov. 2008, pg 1199
[8] Lee, et. al., IEEE NVSMW, 2006, pg. 31
[9] Reid, et. al., IEEE TRED, Oct 2009, pg 2255
[10] Asenov, et. al, IEEE TRED, Apr. 2001, pg. 722
[11] Asenov, et. al., IEEE TRED, May 2003, pg 1254
[12] Kurata, et. al., IEEE JSSC, June 2007, pg. 1362
[13] Bae, et. al. , IEEE TRED, Aug. 2009, pg 1624
[14] Compagnoni, et. al., IEEE TRED, Oct. 2008, 2695
[15] Oh, et. al. , IEEE NVSMW, 2009, pg. 5
[16] Lee, et. al., IEEE NVSMW, 2006, pg. 31
[17] Lee, et. al., IEEE TDMR, March 2004, pg. 110
[18] Kim, et. al, IEEE EDL, 2009, pg 760
[19] Mielke, et. al, IEEE IRPS, 2006, pg. 29.
[20] Belgal, et. al, IEEE IRPS, 2006, pg. 2002, pg. 7
[21] Wang, et. al., IEEE IMW, 2009, pg. 1
[22] Kim, Samsung, Hot Chips Memory Seminar, 2010
[23] Tega, et. al., IEEE IEDM, 2006, pg 1
[24] Pierce, Denali Software , Inc. 2009
[25] Matuoka, IEEE IEDM, 1987, 552
[26] Fukuda, IEEE ISSCC 2011, 11.1
[27] Kim, IEEE IEDM 2010, 1.1
[28] Nikkei Electronics, Feb. 2011, pg. 38
[29] Tanaka, et. al., IEEE VLSI Symp. , 2007 pg. 14
[30] Okano, et. al., APL, 2006 Vol. 76, pg.233
[31] Waser, et. al., Adv. Material, 2009, pg. 2632
[32] Liu, et. al., APL, 2000 Vol. 76, pg. 2749
[33] Li, et. al, IEEE Trans. on Magnetics, Feb. 2005, pg. 909
[34] Kau, et. al., IEEE IEDM, 2009, pg. 617
[35] Wei, et. al. , IEEE IEDM, 2008, pg. 1
[36] S. Raoux, M. Wuttig, Phase Change Materials Science and Applications, Springer 2009
[37] Sosago, et. al., IEEE VLSI 2009, pg. 24
[38] Lee, et. al, IEEE IEDM 2007, 30.2
[39] Park, et. al, Nanotechnology, 2010, Vol. 21, 195201
[40] Gopalakrishnan, et. al, IEEE VLSIT 2010, pg. 205
[41] Wei, et. al, IEEE IEDM 2008
[42] Huang, et. al., Sci. in China, Physics, Mechanics, & Astronomy, 2005, (48), No. 3, pg. 381

Mater. Res. Soc. Symp. Proc. Vol. 1337 © 2011 Materials Research Society
DOI: 10.1557/opl.2011.975

Growth and In-line Characterization of Silicon Nanodots Integrated in Discrete Charge Trapping Non-volatile Memories

J. Amouroux [1,2,3], V. Della Marca [1,2,3], E. Petit [1,2], D. Deleruyelle [2], M. Putero [2], Ch. Muller [2], P. Boivin [1], E. Jalaguier [3], J-P. Colonna [3], P. Maillot [1], and L. Fares [1]

[1] STMicroelectronics, 190 Avenue Célestin Coq, F-13106 Rousset Cedex, France.

[2] Im2np, Institut Matériaux Microélectronique et Nanosciences de Provence, Aix-Marseille Université, Avenue Escadrille Normandie Niemen, F-13397 Marseille Cedex 20, France.

[3] CEA LETI/MINATEC, 17 Rue des Martyrs, F-38054 Grenoble Cedex, France.

ABSTRACT

Non-Volatile Memories (NVM) integrating silicon nanodots (noted SDs) are considered as an emerging solution to extend Flash memories downscaling. In this alternative memory technology, silicon nanocrystals act as discrete traps for injected charges.

Si-dots were grown by Low Pressure Chemical Vapor Deposition (LPCVD) on top of tunnel oxide. Depending on the pre-growth surface treatment, tunnel oxide surface may present either siloxane or silanol groups. SDs deposition relies on a 2–steps process: nucleation by SiH_4 and selective growth with SiH_2Cl_2.

In a context of technological industrialization, it is of primary importance to develop in-line metrology tools dedicated to Si-dots growth process control. Hence, silicon-dots were observed in top view by using an in-line Critical Dimension Scanning Electron Microscopy CDSEM and their average size and density were extracted from image processing. In addition, Haze measurement, generally used for bare silicon surface characterization, was customized to quantify Si-dots deposition uniformity over the wafer. Finally, Haze value was correlated to Si nanodots density and size determined by CDSEM.

INTRODUCTION

Non-Volatile Memories integrating silicon nanodots are considered as an emerging solution for extending Flash memories downscaling. Si-dots are used as discrete traps for injected charges in replacement of the conventional polycrystalline silicon floating gate. For illustration, Fig. 1 shows transmission electron microscopy (TEM) cross sections of Flash cells with either continuous polysilicon floating gate (Fig. 1a) or discrete gate with Si-nanodots (Fig. 1b). Silicon nanodots integration enables improving reliability performances in terms of endurance and retention.

Besides, the robustness to parasitic cross-talk is enhanced and the sensitivity to defects within tunnel oxide is lowered since Si-dots are isolated from each other [1]. Si-nanodots integration is fully compatible with CMOS process and fabrication tools [2]. Since SDs with diameter around 10 nm cover a drastically smaller area than a continuous polysilicon gate, a downscaling toward more aggressive technological node is expected [3,4]. Finally, the lower number of photolithography masks leads to a reduction in wafer cost [5].

This paper shows how silicon nanodots were integrated in Flash-like non-volatile memory cells and describes various in-line and off-line metrology tools dedicated to control and characterize the Si-nanodots growth.

Figure 1. Transmission electron microscopy cross sections of memory stacks integrating either a) standard polycrystalline silicon floating gate or b) discrete gate formed by silicon nanodots.

SILICON NANODOTS GROWTH PROCESS

The first process step consisted in SiO_2 tunnel oxide growth with a thickness two times smaller than one used in standard Flash cells. SiO_2 top surface presents either siloxane (Fig. 2a) or silanol (Fig. 2b) units. In order to increase the number of silanol groups, which are used as nucleation sites for subsequent growth of silicon nanodots, a surface pre-treatment was applied on tunnel oxide [6].

Figure 2. Schematics of a) siloxane and b) silanol units.

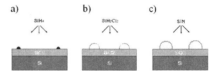

Figure 3. Schematic of Si-nanodots growth process: a) nucleation step by silane SiH_4; b) selective growth by using dichlorosilane (SiH_2Cl_2); c) selective capping by silicon nitride SiN.

Silicon nanodots were grown by Low Pressure Chemical Vapor Deposition (LPCVD) on top of the SiO_2 tunnel oxide. A two steps process was used in either epitaxy tool or in industrial LPCVD furnace. The first step consisted in nucleation with silane (SiH_4) diffusion on hydroxyl tunnel oxide surface (Fig. 3a). In a second step, selective growth was achieved by using dichlorosilane (SiH_2Cl_2) diffusion (Fig. 3b). Thanks to this two step process, Si-nanodots size and density may be controlled independently. Nucleation step enables controlling the number of SDs on top surface of tunnel oxide while the Si-nanodots size is monitored by the selective growth period [7].

SDs growth process was ended with a passivation step (Fig. 3c).This nitridation step that enabled capping the Si-nanodots and preventing parasitic lateral oxidation [8].

MORPHOLOGICAL CHARACTERIZATION

In a context of a "lab-to-fab" technology transfer, it is of primary importance to develop off-line and in-line metrology tools dedicated to the characterization and the control of silicon nanodots. Firstly, to apprehend the morphology and the composition of Si-nanodots, off-line characterization methodologies were developed by using Scanning Electron Microscopy (SEM),

Atomic Force Microscopy (AFM) and Transmission Electron Microscopy (TEM) available in STMicroelectronics Central Laboratory. Besides, in-line metrology was based on Critical Dimension Scanning Electron Microscopy (CDSEM) and Haze measurements.

Off-line characterization

The fundamental technique used to observe silicon nanodots was SEM in either 40° tilted view (Fig. 4a) or top view (Fig. 4b) to determine the distribution of size and density of silicon nanodots. Extraction of Si-nanodots diameter, density and step coverage was achieved thanks to an appropriate image processing performed on SEM top view images.

Atomic Force Microscopy is an alternative technique enabling extraction of information on Si-dots morphology. As shown in Fig. 5, the high resolution of AFM in tapping mode enabled determining dot size, density, spacing between two neighboring dots, silicon dot height, step coverage and growth uniformity over the wafer. Nevertheless, due to the very small size of silicon nanodots size (typically 8-12 nm) and high density (around 10^{12} dots/cm²), AFM was not able to discriminate two adjacent dots even by using high resolution tip. As a consequence, the convolution between sample and tip artificially increases silicon dots size [9].

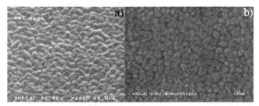 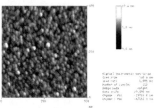

Figure 4. Scanning Electron Microscopy in either a) 40° tilted view or b) top view.

Figure 5. Atomic Force Microscopy used to characterize the morphology of silicon nanodots.

Transmission Electron Microscopy in Ultra High Resolution mode (less than 0.1 nm) was a suited technique to characterize Si-nanodots at atomic scale. After a specific sample preparation by using Focused Ion Beam (FIB), TEM cross sections enabled observing the different layers with the memory stack, confirming the crystalline state of Si-nanodots and determining their crystallographic orientation and height (Fig. 6a). In top view observation, information on Si-dots density, shape and diameter were obtained [7]. Finally, Energy Filtered TEM enabled highlighting of silicon-nanodots (green circle in Fig. 6b). Indeed, this method based on the electron energy loss spectroscopy (EELS) allows filtering the electrons going through the sample according to their kinetic energy. In this mode, Si-nanodots are easily observed but the quality of images is conditioned by the sample preparation which is very long and hard to perform.

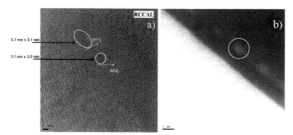

Figure 6. Silicon nanodots Transmission Electron Microscopy cross sections obtained either a) on Ultra High Resolution mode, or b) by using energy filtering.

In-line characterization

Si-nanodots were observed after growth process by using in-line top view Critical Dimension Scanning Electron Microscopy (CDSEM). This characterization technique appears as the most promising to observe nanodots but requires some dedicated patterns to focus of electron beam on the sample. For various SDs size and density, accurate observations were done (Fig. 7). In complement, extraction of SDs diameter and density was done by appropriate image processing.

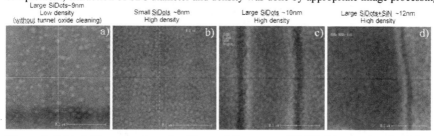

Figure 7. Critical Dimension Scanning Electron Microscopy (CDSEM) in top view for different distributions of Si-nanodots: a) large dots with very low density (no surface treatment of tunnel oxide); b) small dots and high density; c) large dots and high density; d) high density of large dots capped with silicon nitride.

In addition, Haze measurements, generally used for bare silicon surface characterization (KLA-Tencor SP1™), were customized to quantify Si-nanodots deposition uniformity over the wafer. In the conventional method, laser beam with a 488 nm wavelength is scattered by defects distributed over the different wafer surface. Haze value corresponds to a measurement of the light scattered off of the wafer surface: lower roughness translates into less scattered light and a lower haze. Figure 8 shows typical Haze mappings measured on Si-nanodots grown either in an industrial LPCVD furnace (Fig. 8a) or in an epitaxy tool (Fig. 8b): it is clearly seen that a better uniformity is achieved on LPCVD furnace. Some attempts were done to correlate Haze value to Si-dots size and density determined by CD-SEM. Figure 8c shows that the dot size increases and their density decreases along with Haze value. The main goal of such a methodology is to build an abacus for various Si-dots growth process parameters to link a SDs recipe to a Haze value.

This chart will ultimately use SP1™ as a fast and efficient Si-nanodots growth process control tool.

a)   b)   c)

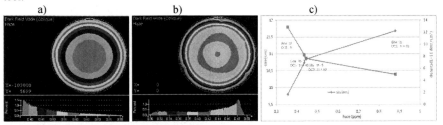

Figure 8. Haze mapping of Si-nanodots growth uniformity of wafers processed either a) in an industrial LPCVD furnace or b) in epitaxy tool. c) Evolution of Si-dots size and density extracted from CD-SEM as a function of Haze value.

ELECTRICAL CHARACTERIZATION

Electrical characterization was performed on Flash SDs-based cells consisting of ONO inter-poly dielectric with equivalent thickness oxide of 14 nm, Si-nanodots of 8-10 nm in diameter and capped by 2 nm of SiN. A 5.5 nm tunnel oxide separated SDs from silicon P-substrate. Geometrical cell dimensions were 70 nm in width and 140 nm in length. A standard experimental setup was used with a pulse generator HP8110 and a parameter analyzer HP4156.

Figure 9 shows programming and erasing characteristics obtained with different gate voltage biases (V_p/V_e). The cell was programmed by channel hot electron (CHE) using a fixed 4.2V drain voltage (V_d) and was erased by Fowler Nordheim (FN) tunneling. The amplitude of programming window (ΔV_{th}) was calculated as the difference between programmed (Vth_{prg}) and erased (Vth_{ers}) threshold voltages. The programming window seems to saturate for high program time contrary to erasing window. Fig. 10 shows endurance characteristics measured up to 10^5 cycles. It is clearly observed a V_{th} shift in both states due to the parasitic charge trapping in ONO layers [10]. Programming and erasing conditions progressively shrink the programming window from 5V to 3V after 10^5 cycles. The two states remains separated with a difference of 1.6V between Vth_{prg} at the endurance test beginning and Vth_{ers} at the end. Hence, encouraging results were obtained, in particular a 5 V programming window reached with a fast programming operation of 10µs.

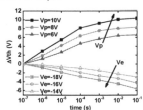

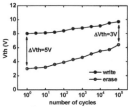

Figure 9. Programming (CHE) and erasing (FN) characteristics with different gate voltage (Vp/Ve). $\Delta V_{th}=Vth_{prg} - Vth_{ers}$.

Figure 10. 10^5 cycles endurance characteristic. Programming: $V_p=10V$, $V_d=4V$, $V_s=V_b=0V$, $t_p=10µs$; Erasing: $V_e=-18V$, $V_d=V_s=V_b=0V$, $t_e=50ms$.

19

CONCLUSIONS

This paper demonstrates the successful integration of silicon nanodots in Flash-like memory cells. The present experimental conditions led to larger nanodots as compared to data generally published in literature (8-10 nm *versus* 5 nm). The surface treatment preceding Si-dot growth increases the number of nucleation sites and certainly enhances the coalescence process and enlarges to improve wafer surface covering (larger dots enable increasing stored charge). Beside process challenges that manufacturing is facing, in-line and off-line metrology is mandatory to control the nanodots integration in high performances non-volatile memories. Regarding in-line characterization, CDSEM is surely the most suited technique to determine SDs size and density. Measurement automation and routine image processing are in progress. Besides, SP1^{TM} appears as a promising method to link growth recipes to morphological characteristics on Si-nanodots.

Concerning off-line techniques, SEM enables effortless observing Si-dots morphology and extracting size, density and step coverage. In complement, despite a precise and time-consuming sample preparation, TEM in either high resolution mode or in energy filtering mode enables accessing information on Si-nanodots at atomic scale (*e.g.* crystallographic orientation).

Finally, encouraging electrical results were obtained on Flash-like memory cells integrating silicon nanodots. At present, optimization of memory stack and Si-dots growth recipe tuning are in progress to further improve the electrical performances especially in terms of endurance.

ACKNOWLEDGMENTS

The authors would like to thank Jean-Philippe Colonna (CEA-LETI), Laurent Brun, Frederic Guemas, Benoist Richard, Antonello Scanni (STMicroelectronics) for process and Si-Dots integration in memory stack; Manfred Regula, Alexandre Vauselle, Valerie Reboul, Luigi Verderio (STMicroelectronics), Sylvie Favier, and Emmanuel Nolot (CEA-LETI) for in-line characterization and Rousset Central Characterization Analysis Laboratory for off-line morphological and electrical characterizations.

REFERENCES

1. B. De Salvo *et al.*, *IEDM Tech. Dig.*, 597 (2003).
2. C. Monzio Compagnoni, D. Ielmini, A.S. Spinelli, A.L. Lacaita, C. Gerardi, L. Perniola, B. De Salvo, and S. Lombardo, *IEDM Tech. Dig.*, 549 (2003).
3. G. Molas, D. Deleruyelle, B. De Salvo, G. Ghibaudo, M. Gely, S. Jacob, D. Lafond, and S. Deleonibus, *IEDM Tech. Dig.*, 877 (2004).
4. J. Brault, M. Saitoh, and T. Hiramoto, *IEEE Trans. Nano.* **4**, 349 (2005).
5. R. Muralidhar *et al.*, *IEDM Tech. Dig.*, 601 (2003).
6. F. Mazen, T. Baron, G. Brémond, N. Buffet, N. Rochat, P. Mur, and M.N. Séméria, *J. Electrochem. Soc.* **150**, G203 (2003).
7. F. Mazen, T. Baron, A.M. Papon, R. Truche, J.M. Hartmann, *Appl. Surf. Sci.* **214**, 359 (2003).
8. J-P. Colonna, G. Molas, M.W. Gely, M. Bocquet, E. Jalaguier, B. De Slavo, H. Grampeix, P. Brianceau, K. Yckache, A-M. Papon, G. Auvert, C. Bongiorno, and S. Lombardo, *Mater. Res. Soc. Symp. Proc.* **1071**, 1071-F02-02 (2008).
9. T. Baron, F. Martin, P. Mur, C. Wyon, M. Dupuy, C. Busseret, A. Souifi, and G. Guillot, *Appl. Surf. Sci.* **164**, 29 (2000).
10. R.F. Steimle *et al.*, Jr., *IEEE Proc. Non Volatile Semicond. Memory Workshop*, 73 (2004).

Mater. Res. Soc. Symp. Proc. Vol. 1337 © 2011 Materials Research Society
DOI: 10.1557/opl.2011.976

Matrix Density Effect on Morphology of Germanium Nanocrystals Embedded in Silicon Dioxide Thin Films

Arif S. Alagoz[1, 2], Mustafa F. Genisel[3, 4], Steinar Foss[5], Terje G. Finstad[5], Rasit Turan[2]

[1] Department of Applied Science, University of Arkansas at Little Rock, Little Rock, AR, U.S.A
[2] Department of Physics, Middle East Technical University, Ankara, Turkey.
[3] Department of Chemistry, Middle East Technical University, Ankara, Turkey.
[4] Department of Chemistry, Bilkent University, Ankara, Turkey.
[5] Department of Physics, University of Oslo, Oslo, Norway.

ABSTRACT

Flash type electronic memories are the preferred format in code storage at complex programs running on fast processors and larger media files in portable electronics due to fast write/read operations, long rewrite life, high density and low cost of fabrication. Scaling limitations of top-down fabrication approaches can be overcome in next generation flash memories by replacing continuous floating gate with array of nanocrystals. Germanium (Ge) is a good candidate for nanocrystal based flash memories due its small band gap. In this work, we present effect of silicon dioxide (SiO_2) host matrix density on Ge nanocrystals morphology. Low density Ge+SiO_2 layers are deposited between high density SiO_2 layers by using off-angle magnetron sputter deposition. After high temperature post-annealing, faceted and elongated Ge nanocrystals formation is observed in low density layers. Effects of Ge concentration and annealing temperature on nanocrystal morphology and mean size were investigated by using transmission electron microscopy. Positive correlation between stress development and nanocrystal size is observed at Raman spectroscopy measurements. We concluded that non-uniform stress distribution on nanocrystals during growth is responsible from faceted and elongated nanocrystal morphology.

INTRODUCTION

One of the main issues in information technology (IT) is the storage of digital information. Nanocrystal based new generation flash memories are promising to follow Moore's law and overcome scaling limitations with higher operation performance. Intensive researches have been focused especially on germanium nanocrystal [1-9] due to its fabrication compatibility with current complementary metal oxide semiconductor (CMOS) technology and small band gap, providing short writing/erasing and long retention time [2,3]. Ion implantation [4], chemical vapor deposition [5], laser ablation deposition [6] and magnetron sputter deposition [7-9] have been widely used techniques to fabricate germanium nanocrystals embedded in a various host matrix.

In this study, we fabricated germanium nanocrystals embedded in silicon dioxide host matrix by off-angle magnetron sputter deposition and high temperature post annealing. We sandwiched low density Ge+SiO_2 layers between higher density SiO_2 layers. After post annealing, we observed non-spherical and faceted Ge nanocrystals in low density Ge+SiO_2 layers. TEM measurements showed that atomic Ge concentration, annealing temperature and

SiO$_2$ matrix density are the critical parameters for nanocrystals' mean size and morphology. Raman spectroscopy measurements also showed correlation between stress formation and Ge nanocrystals size after annealing. Non-uniform matrix density is addressed as the main reason of non-uniform stress distribution on nanocrystals which in turn cause non-spherical nanocrystal forms.

EXPERIMENT

N-type Si (100) with 40 nm thermal oxide wafers were subjected to RCA I & II surface cleaning procedures. Samples were loaded to Vaksis nano-D 100 magnetron sputter deposition system and chamber pumped down to 7×10^{-7} Torr before each deposition. Periodic Ge+SiO$_2$/SiO$_2$ thin films were deposited at room temperature by sputtering 3 inch diameter Ge and SiO$_2$ targets with Argon plasma. SiO$_2$ and Ge targets were located at $+25°$ and $-25°$ off-angle with substrate normal, respectively. Sputtering pressure was fixed at 3 mTorr during depositions. RF power applied to SiO$_2$ target was fixed and Ge concentration of each layer was controlled by DC power applied to the Ge target. Thicknesses of the films were controlled by monitoring thickness monitor and deposition time, details of process parameters are listed in Table 1. Samples were annealed in a fused quartz furnace at 600°C, 700°C, 750°C and 800°C temperatures under vacuum (3.4×10^{-5} Torr) for 30 minutes to form Ge nanocrystals. After post annealing, cross sectional transmission electron microscopy (TEM) measurements were performed in order to investigate nanocrystals' size and morphology. Each Ge+SiO$_2$/SiO$_2$ periodic layer of multilayer sample was deposited on another set of substrates with the same deposition conditions and subjected to the same annealing conditions. Samples were named as Ge-L2, Ge-L3, Ge-L4 and Ge-L5 referring to each corresponding layer of multilayer sample (see Table 1). Raman spectroscopy measurements are conducted on these samples to analyze Ge crystallization and stress formation after annealing.

Table 1 Deposition parameters of multilayer and corresponding two layer samples

Layer #	Sample name	Ge target power (W)	SiO$_2$ target power (W)	Deposition time (min)
5	Ge-L5	4	175	24
4	Ge-L4	11	175	20
3	Ge-L3	22	175	14
2	Ge-L2	55	175	8
1	-	66	0	10

Cross sectional transmission electron microscope measurements were carried out by analytical JEOL2000FX TEM at 200 keV. Backscattering Raman measurements were performed by Jobin Yvon Horiba confocal micro-Raman at room temperature. He-Ne laser (632.83 nm) was used as an excitation source, double monochromator and Peltier cooled CCD detector were used to detect Raman shifts.

DISCUSSION

As shown in figure 1a, there is no indication of Ge nanocrystal formation in any layer of as-sputtered multilayer sample. After post annealing, layer 1 (%100 Ge) turns into a polycrystalline thin film. Nanocrystal formation starts after annealing even at 600°C in layers 2, 3 and 4 as shown in figure 1b. Due to high Ge concentration in layer 2, nanocrystals' size increases dramatically with increasing annealing temperature and even exceeds the co-sputtered layer thickness; in addition nanocrystals in this layer took faceted and elongated forms. In figure 1b-e, well separated nanocrystals can be observed in layer 3 after annealing. Mean nanocrystal size in this layer is comparable with deposited layer thickness and again non-spherical faceted and elongated nanocrystals observed. Well separated and uniformly distributed nanocrystals are observed in layer 4. In this layer, nanocrystals' mean size increases smoothly with increasing annealing temperature and the size variation is smaller. On the other hand, no crystallization can be detected in layer 5 due to the low Ge concentration.

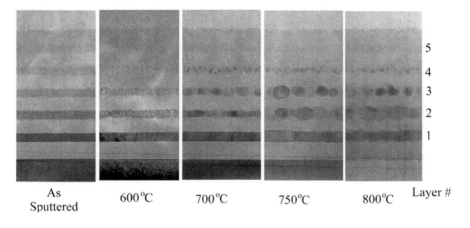

| As Sputtered | 600°C | 700°C | 750°C | 800°C | Layer # |

50 nm

Figure 1 Cross sectional transmission electron microscope images of multilayer Ge+SiO$_2$ as-sputtered and post-annealed samples at 600°C, 700°C, 750°C, 800°C for 30 min.

Non-uniform compressive stress on nanocrystals formed by SiO$_2$ host matrix can be addressed as the reason of faceted and elongated nanocrystal morphology [10]. ±25° off-angle co-sputtering of Ge and SiO$_2$ results in non-uniform low density film which causes non-uniform stress after annealing. In layer 2 and 3, higher density SiO$_2$ layers behaves as barriers during nanocrystal growth, hence Ge nanocrystals preferentially elongate along co-sputtered layers.

Correlation between Ge nanocrystal formation and stress built-up on nanocrystals were investigated by Raman spectroscopy in Ge-L2, Ge-L3, Ge-L4 and Ge-L5 samples. As expected, as-sputtered and Ge-L5 samples did not show any indication of nanocrystal formation due to low deposition temperature and low Ge concentration, respectively. Weak Ge-Ge transverse optical

(TO) Raman signal of Ge-L4 lost in the broad longitudinal optical (LO) peak of Si substrate centered at 301 cm^{-1} [11], therefore nanocrystal formation couldn't observed. Figure 2a shows Ge-Ge TO peak at 300 cm^{-1} after annealing of sample Ge-L3 at 600°C. Ge-Ge TO peak intensity increases gradually at 700°C and 750°C then decreases at 800°C. Note that Si LO peak is fixed at 301 cm^{-1} while Ge-Ge TO signal shifts with increasing annealing temperature. Correlation of Ge-Ge TO peak position and nanocrystals' mean size as a function of annealing temperature is shown in figure 2b.

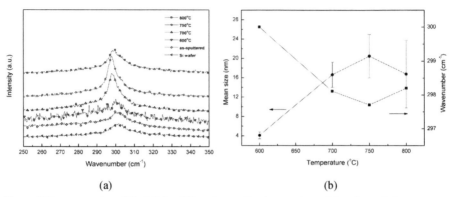

(a)   (b)

Figure 2 Raman spectra of Ge-L3 for different annealing temperatures (a) and correlation between Ge-Ge TO peak shift and mean nanocrystal size with changing post-annealing temperature.

It is know that Ge-Ge TO peak shift to high wavenumbers with increasing nanocrystal size due to phonon confinement effect [12, 13] and shift to low wavenumbers is due to the stress formation on the nanocrystals [14, 15]. Although, nanocrystal size increase and stress formation are coupled latter dominates in these samples. This strong relationship also confirms high compressive silicon dioxide host matrix stress induced on germanium nanocrystals.

CONCLUSIONS

We fabricated germanium nanocrystals in low density Ge+SiO$_2$ layer between higher density SiO$_2$ layers by using off-angle magnetron sputter deposition and high temperature post annealing. We observed non-spherical and faceted Ge nanocrystals in low density Ge+SiO$_2$ layers and attributed this effect to non-uniform compressive stress formation in non-uniform low density host matrix. We found that atomic Ge concentration, annealing temperature and SiO$_2$ matrix density are the critical parameters for nanocrystals' mean size and morphology. Raman spectroscopy confirmed germanium nanosrystal formation and high compressive silicon dioxide stress induced on germanium nanocrystals.

ACKNOWLEDGMENTS

This work has been partially supported by the EU FP6 projects SEMINANO under the Contract No. NMP4 CT2004 505285.

REFERENCES

1. H. I. Hanafi, S. Tiwari and I. Khan IEEE T. Electron Dev. **43** 1553 (1996)
2. M. She and T. J. King IEEE T. Electron Dev. **50** 1934 (2003)
3. V. Beyer, J. von Borany and M. Klimenkov J. Appl. Phys. **101** 094507 (2007)
4. E. S. Marstein, A. E. Gunnæs, U. Serincan, S. Jørgensen, A. Olsen, R. Turan and T. G. Finstad Nucl. Instrum. Meth. B **207** 424 (2003)
5. S. Ağan, A. Çelik-Aktas, J. M. Zuo, A. Dana and A. Aydınlı Appl. Phys. A-Mater. **83** 107 (2006)
6. W. L. Liu, P. F. Lee, J. Y. Dai, J. Wang, H. L. W. Chan, C. L. Choy, Z. T. Song and S. L. Feng Appl. Phys. Lett. **86** 013110 (2005)
7. N. A. P. Mogaddam, A. S. Alagoz, S. Yerci, R. Turan, S. Foss and T. Finstad J. Appl. Phys., **104**, 124309 (2008)
8. P. Basa, A. S. Alagoz, T. Lohner, M. Kulakci, R. Turan, K. Nagy and Zs. J. Horváth, App. Surf. Sci., **254**, 3626 (2008)
9. A. Gencer Imer, S. Yerci, A. S. Alagoz, M. Kulakci, U. Serincan, T. G. Finstad and R. Turan J. of Nano. and Nanotech.**10**, 525 (2010)
10. A. V. Kolobov, S. Q. Wei, W. S. Yan, H. Oyanagi, Y. Maeda and K. Tanaka Phys. Rev. B **67** 195314 (2003)
11. A. V. Kolobov J. Appl. Phys. **87** 2926 (2000)
12. U. Serincan, G. Kartopu, A. Guennes, T. G. Finstad, R. Turan, Y. Ekinci and S. C. Bayliss Semicond. Sci. Tech. **19** 247 (2004)
13. M. Fujii, S. Hayashi and K. Yamamoto Jpn. J. Appl. Phys. **30** 687 (1991)
14. W. K. Choi, H. G. Chew, F. Zheng, W. K. Chim, Y. L. Foo and E. A. Fitzgerald E A Appl. Phys. Lett. **89** 113126 (2006)
15. I. D. Sharp, D. O. Yi, Q. Xu, C. Y. Liao, J. W. Beeman, Z. Liliental-Weber, K. M. Yu, D. N. Zakharov, J. W. Ager III, D. C. Chrzan and E. E. Haller Appl. Phys. Lett. **86** 063107 (2005)

Mater. Res. Soc. Symp. Proc. Vol. 1337 © 2011 Materials Research Society
DOI: 10.1557/opl.2011.1068

Temperature Effects on Charge Transfer Mechanisms of nc-ITO Embedded ZrHfO High-*k* Nonvolatile Memory Devices

Chia-Han Yang[1,2], Yue Kuo[1], Chen-Han Lin[1] and Way Kuo[3]

[1]Thin Film Nano & Microelectronics Research Laboratory, Texas A&M University,
College Station, TX 77843-3122, U.S.A.
[2]Department of Industrial and Information Engineering, University of Tennessee,
Knoxville, TN 37996, U.S.A.
[3]City University of Hong Kong, Hong Kong

ABSTRACT

The nanocrystalline ITO embedded Zr-doped HfO_2 high-k dielectric thin film has been made into MOS capacitors for nonvolatile memory studies. The devices showed large charge storage densities, large memory windows, and long charge retention times. In this paper, authors investigated the temperature effect on the charge transport and reliability of this kind of device in the range of 25°C to 125°C. The memory window increased with the increase of the temperature. The temperature influenced the trap and detrap of not only the deeply-trapped but also the loosely-trapped charges. The device lost its charge retention capability with the increase of the temperature. The Schottky emission relationship fitted the device in the positive gate voltage region. However, the Frenkel-Poole mechanism was suitable in the negative gate voltage region.

INTRODUCTION

Silicon dioxide (SiO_2) has been used as the gate dielectric of MOS devices for decades. As its thickness is reduced from 3.5 nm to 1.5 nm, the leakage current density increases drastically, e.g., from 10^{-12} A/cm^2 to 10 A/cm^2 at a gate bias of 1 V, due to quantum-mechanical tunneling [1]. Currently, there are many researches on replacing SiO_2 with a high dielectric constant (high-k) material, such as Si_3N_4, $HfSi_xO_y$, HfO_2, and ZrO_2, to achieve a low leakage current at the same time to improve the device performance and reliability [2]. High-k dielectrics are also critical for nanosize nonvolatile memory (NVM) devices [1]. Recently, the nanocrystals embedded high-k dielectric structure has been proposed to replace the conventional polysilicon floating gate structure for the nonvolatile memory application for advantages of the low operating power and improvement of reliability [3-5]. Due to the low band offset between the high-k film material and silicon, this kind of device requires a low operating power. Therefore, the nanocrystals embedded high-k structure is an attractive structure for the high-density NVM. However, conventional high-k materials such as ZrO_2 or HfO_2 usually crystallize at a low temperature, e.g., <600°C [5]. The Zr-doped HfO_2 (ZrHfO), on the other hand, has many advantages over the undoped HfO_2 in areas such as the crystallization temperature, the effective *k* value, and the interface state density [6-9]. Recently, nanocrystalline (nc) metals, metal oxides or semiconductor materials, such as Ru, Ni, indium tin oxide (ITO), Si, and zinc oxide (ZnO), have been embedded into the high-k dielectric for the NVMs applications [10-14]. For example, the energy band diagram of the nc-ITO embedded ZrHfO was published previously [11]. Based on the band offsets between nc-ITO and Si, the nc-ITO embedded high-k gate dielectric should have excellent charge retention characteristics [14]. It has been demonstrated that the ZrHfO/nc-ITO/ZrHfO tri-layer structure can trap a large number of charges with a long retention time [14]. Yang et al. [15] investigated the stress-induced deterioration of the nc-ITO embedded ZrHfO film. The deterioration of this kind of capacitor was observed from the decrease of the capacitance in the accumulation region and the shift of flatband voltage as well as the change of the amount of induced charges. However, most reliability studies on nonvolatile memories are done at the room temperature [15-16]. In the accelerated life test, it is necessary to understand how the device performs at the raised temperature. In this paper, authors investigated the temperature influence on the retention characteristics and the charge transportation mechanism of the nc-ITO embedded ZrHfO MOS capacitor.

EXPERIMENT

The MOS capacitor based on the ZrHfO/nc-ITO/ZrHfO tri-layer structure has been prepared, as shown in Figure 1. The bottom tunnel oxide, nc-ITO, and top control oxide layers were sputter deposited sequentially in the same chamber with a one pumpdown process without breaking the vacuum. The HF pre-cleaned p-type Si (100) wafer (doping concentration at 10^{15} cm^{-3}) was used as the substrate. Both the tunnel and the control ZrHfO films were deposited from the Hf/Zr (88:12 wt%) composite target in an Ar/O$_2$ (1:1) mixture at 5 mTorr at 60W and room temperature, e.g., 2 min for the former and 10 min for the latter. The ITO film was sputter deposited from the ITO (In/Sn:90/10 wt%) target in Ar/O$_2$ (1:1) at 10 mTorr and 80W for 30 sec. The control sample, i.e., containing only the ZrHfO gate dielectric without the embedded ITO layer, was also prepared and characterized for comparison. The as-deposited amorphous ITO layer was transformed into discrete nanocrystals after post-deposition annealing (PDA) at 800°C for 60 sec under N$_2$ by rapid thermal annealing. An aluminum (Al) film was sputtered deposited on top of the high-k stack and subsequently wet etched to form the gate pattern with an area of 7.85×10^{-5} cm^2. The backside of the wafer was deposited with Al for ohmic contact. The complete capacitor, i.e., with the aluminum gate electrode and back contact, was annealed at 300°C under H$_2$/N$_2$ (90/10) for 5 minutes. Lin and Kuo [14] reported that the 3-5 nm size, (222) oriented nc-ITO dots are formed in the ZrHfO matrix after PDA. Discrete nc-ITO dots, as highlighted in circles in Fig. 2, are observable in the cross-sectional TEM picture [14]. The equivalent oxide thickness (EOT) of the nc-ITO embedded high-k stack was 7.8 nm estimated from the C-V curve at 1MHz.

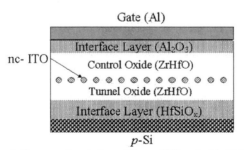

Figure 1. Cross-sectional view of the nc-ITO embedded MOS capacitor

28

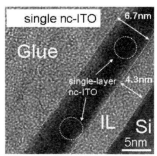

Fig. 2. Cross-sectional TEM view of the nc-ITO embedded ZrHfO film [14].

DISCUSSION

The temperature effect on the charge retention efficiency of the nc-ITO embedded capacitor was studied with the following method [10-11]. First, a gate bias (V_g) was applied to the capacitor for a period of time to trap charges to the dielectric structure. Second, after releasing the V_g, the C-V curve was measured in a small V_g range, i.e., -2V to +1V. Third, the C-V measurement step was repeated every 1800s. Only negligible charges were injected into or removed from the capacitor during the C-V measurement because of the small V_g range. The flat band voltage, V_{FB}, was calculated from the C-V curve to reflect the capacitor's charging state. The flat band voltage shift (ΔV_{FB}), which is defined as V_{FB} (after stress release for a period of time) - V_{FB} (before stress), is expressed as a function of the time (t), as shown in Figure 3(a) after releasing the 1s V_g = -8V stress conditions at different temperatures. First, the magnitude of the memory window increases with the increase of the temperature, i.e., 0.64V, 1.14V, and 1.21V at 25°C, 75°C, and 125°C, respectively. In contrast, the control sample has negligible memory windows, e.g., ΔV_{FB} =0.03V and 0.3V at 25°C and 75°C, respectively under the same sweeping conditions. All trapped charges in the control sample were lost after 28 hours at 75°C. The initial increase of the memory window in the nc-ITO embedded sample is mainly due to the change in the Fermi level and the interface properties [17]. Compared with the room temperature stress condition, the elevated temperature provides the additional energy to holes to overcome the barrier height for easier reaching the nc-ITO site. The ΔV_{FB} - t curves in Fig. 3(a) is consisted of two sections: quick loss of charges at t < 3600 seconds followed by the slow loss of the remaining charges after a long period of time. The first section is due to the detraping of the loosely trapped charges; the second section is due to the release of the strongly trapped charges. The initial charge decay rate and the loss of the trapped charges increase with the increase of temperature, i.e., 27.6%, 45.7% and 71.5% at 25°C, 75°C and 125°C, respectively after one hour. After 10 hours, the capacitor lost 34.8%, 64% and 92% of the total trapped charges at 25°C, 75°C and 125°C, respectively. This is because the conductance of the dielectric material increases with the increase of the temperature, which facilitates the leakage of the charges to the Si wafer. The increase of the leakage current with temperature has been observed previously [18]. Therefore, temperature not only influences those loosely-trapped but also strongly-trapped charges. Figure 3(b) shows the same data as Fig. 3(a) but with the time on the logarithm scale extrapolated lifetime. At 25°C, 61% of the stored charges were lost after 10 years. However, at high temperatures, the loss of the charge increased drastically. For example, at 75°C, all stored charges were lost after 270 days and at 125°C the time was shortened to 1.67 days.

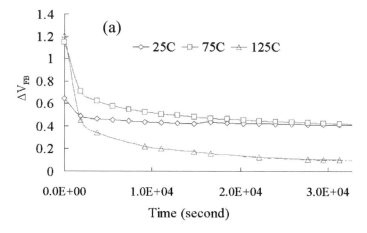

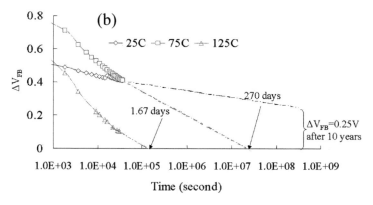

Figure 3. (a) Charge retention characteristic of nc-ITO embedded high-k thin film under different temperatures, (b) same as (a) with time on the logarithm scale

The thermal effect of the charge transport mechanism of the nc-ITO embedded sample was also studied. Figure 4 shows the J–V curves of the nc-ITO embedded capacitor under different temperatures in the log-lin scale. The gate voltage was swept from 0 to +6V in Fig. 4(a) and 0 to -6V in Fig. 4(b). Above +2V, an electron-rich inversion layer is formed; below -3V, a hole accumulation layer is established. Under the positive V_g bias condition, the leakage current is contributed by the transport of the minority carriers through the high-k stack. However, when the bias positive V_g is very large, almost all electrons in the inversion layer are deleted. The inversion layer is near intrinsic. The leakage current cannot increase further with the increase of the bias V_g. Therefore, a plateau is formed in the I-V curve. Since the concentration of the carriers in the intrinsic Si increases with the increase of temperature, more electrons exist in the inversion region at the high temperature than at the low temperature. The magnitude of the leakage current increases with the increase of temperature. Therefore, the onset voltage of the plateau in the I-V curve increases with the increase of temperature. Fig. 4(a) shows that the onset V_g's of the plateaus in the I-V curves are 1.9V, 2.2V and 2.3V at 25°C, 75°C and 125°C, respectively. Both figures show that the leakage current density increases with the increase of temperature, which is consistent with the Frenkel-Poole or Schottky emission mechanism. In addition, the magnitude of the leakage current under negative bias is smaller than that of the positive bias. Since the barrier height of the hole between HfSiO$_x$ and Si is larger than that of the electron, i.e., 3.4 eV vs. 1.5 eV, the leakage current density in Fig. 4(a) is higher than that in Fig. 4(b) under the same magnitude of V_g. Previously, it was reported that the charge transports through the HfO$_2$ film following the Frenkel-Poole (F-P) mechanism or the Schottky emission [19-20], as shown in equation (1) and (2), respectively.

$$J \sim E \times \exp\left[\frac{-q\left(\Phi_B - \sqrt{qE/\pi\varepsilon}\right)}{kT}\right] \qquad (1)$$

$$J = A \times T^2 \exp\left[\frac{-q\left(\Phi_B - \sqrt{qE/\pi\varepsilon}\right)}{kT}\right] \qquad (2)$$

where E is the electric field, T is the temperature, ε is the insulator dynamic permittivity, m is the effective mass, Φ_B is the barrier height, and q is the charge of one electron. From the above equations, Schottky emission mechanism is more dependent on the temperature than F-P conduction mechanism. Figure 4 (a) shows that the Schottky emission is suitable for the leakage of current under the positive V_g bias condition. Figure 4(b) shows that the F-P conduction mechanism is applicable for the negative V_g bias condition. The $\ln(J/T^2)$ vs. $E^{1/2}$ curves (Schottky emission fitting plot) and $\ln(J/E)$ vs. $E^{1/2}$ curves (F-P fitting plot) are shown in Figure 5. The E value was calculated from V_g and the physical thickness of the high-k stack (from the TEM picture in Fig. 2). The result shows that the onset of the Schottky emission dominates the charge transportation mechanism at $V_g > 2.5$V. Since the Si/HfSiO$_x$ conduction band offset is much smaller than the valence band offset, i.e., 1.5V vs. 3.4V, the Schottky emission mechanism is more pronounced than the F-P conduction mechanism under positive V_g [19]. However, under the negative V_g, the Schottky emission mechanism is overwhelmed by the F-P conduction due to the large valence band offset between Si and HfSiO$_x$.

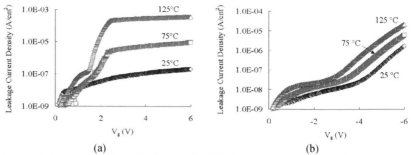

(a)   (b)

Figure 4. J-V curve of the nc-ITO embedded MOS capacitor measured (a) from 0V to +6V and (b) from 0V to -6V

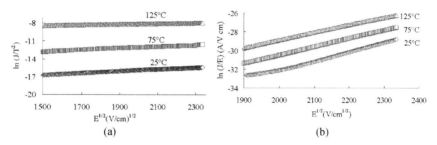

(a)   (b)

Figure 5. (a) Fitting of Schottky relationship under the positive V_g condition and (b) fitting of the Frenkel-Poole relationship under the negative V_g condition at 25°C, 75°C, and 125°C.

CONCLUSIONS

Temperature effects on the charge retention characteristics and transportation mechanism of the nc-ITO embedded high-k dielectric have been investigated. The initial memory window increased with the increase of temperature. However, the charge storage capacity decreased with the increase of temperature. After releasing the stress for 10 hours, the loss of the trapped charges were 34.8%, 64% and 92% at 25°C, 75°C and 125°C, respectively due to the increase of thermal energy of the trapped charge and the electric conductivity of the high-k film. The charge retention capability decreased with the increase of temperature, i.e., $\Delta V_{FB} = 0.25V$ after 10 years at 25°C, 0V after 270 days at 75°C, and 1.67 days at 125°C. The charge transport through the high-k stack follows the Schottky emission relationship in the positive V_g range; it follows the trap-assisted F-P mechanism in the negative V_g range. However, the exact charge transport mechanism in the former needs further studies.

ACKNOWLEDGMENTS

This research is supported by NSF CMMI Grant No. 0926379.

REFERENCES

1. The International Technology Roadmap for Semiconductors. Semiconductor Industry Association, December 2003.
2. S. Chatterjee, S. K. Samanta, H. D. Banerjee and C. K. Maiti, in *Semicond. Sci. Technol.*, vol. 17, p. 993, 2003.
3. S. Tiwari, F. Rana, H. Hanafi, A. Hartstein, E. F. Crabbé, and K. Chan, in *Appl. Phys. Lett.*, vol. 68, no. 10, p. 1377, 1996.
4. J. J. Lee and D. L. Kwong, *IEEE Trans. Electron Devices*, vol. 52, p. 507, 2005.
5. J. Lu, Y. Kuo, J. Yan, and C.-H. Lin, *Jpn. J. Appl. Phys.*, vol. 45, L901, 2006.
6. Y. Kuo, J. Lu, S. Chatterjee, J. Yan , T. Yuan , H.- C. Kim, W. Luo, J. Peterson and M. Gardner, in *ECS Trans*, vol. 1, no. 5, p. 447, 2006.
7. D. H. Triyoso, in *ECS Trans.*, vol. 3, no. 3, p. 463, 2006.
8. Y. Kuo, in *ECS Trans.*, vol. 3, no. 3, p. 253, 2006.
9. Y. Kuo, in *ECS Trans.*, vol. 2, no. 1, p. 13, 2006.
10. J. Lu, C.-H. Lin and Y. Kuo, in *JES*, vol. 115, no. 6, H386, 2008.
11. A. Birge and Y. Kuo, in *JES*, vol. 154, no. 10, H887, 2007.
12. D. B. Farmer and R. G. Gordon, in *J. Appl. Phys.*, vol. 101, 124503, 2006.
13. J. J. Lee, Y. Harada, J. W. Pyun and D. L. Kwong, in *Appl. Phys. Lett.*, vol. 86, 103505, 2005.
14. C. H. Lin and Y. Kuo, *ECS Trans.*, vol. 28, no. 1, p. 269, 2010.
15. C. H. Yang, Y. Kuo, C. H. Lin, and W. Kuo in *ECS Trans.*, vol. 33, no. 3, p. 307, 2010.
16. C. H. Yang, Y. Kuo, C. H. Lin, and W. Kuo in *ECS Trans.*, vol. 25, no. 6, p. 457, 2009.
17. M. H. Weng, R. Mahapatra, P. Tappin, B. Miao, S. Chattopadhyay, A.B. Horsfall and N.G. Wright, *Materials Sci. Semicond. Proc.* vol. 9, p. 1133, 2006.
18. C. H. Yang, Y. Kuo, C. H. Lin, and W. Kuo, in *Mat. Res. Soc. Symp. Proc.*, vol. 1160, H02, 2009.
19. W. J. Zhu, T. P. Ma, T. Tamagawa, J. Kim, and Y. Di, in *IEEE Electron Device Letters*, vol. 23 no. 2, p. 97, 2002.
20. Z. Xu, M. Houssa, S. De Gendt and M. Heyns, *in Appl. Phys. Lett.*, vol. 80. no. 11, p. 1975, 2002.

Mater. Res. Soc. Symp. Proc. Vol. 1337 © 2011 Materials Research Society
DOI: 10.1557/opl.2011.1069

Enhancement of Nonvolatile Floating Gate Memory Devices Containing AgInSbTe-SiO$_2$ Nanocomposite by Inserting HfO$_2$/SiO$_2$ Blocking Oxide Layer

Kuo-Chang Chiang, Tsung-Eong Hsieh

Department of Materials Science and Engineering, National Chiao Tung University,

1001 Ta-Hseuh Road, Hsinchu, Taiwan 30010, R.O.C.

ABSTRACT

This work presents an enhancement of nonvolatile floating gate memory (NFGM) devices comprised of AgInSbTe (AIST) nanocomposite as the charge-storage trap layer and HfO$_2$ or HfO$_2$/SiO$_2$ as the blocking oxide layer. A significantly large memory window (ΔV_{FB}) shift = 30.7 V and storage charge density = 2.3×10^{13} cm^{-2} at ±23V gate voltage sweep were achieved in HfO$_2$/SiO$_2$/AIST sample. Retention time analysis observed a ΔV_{FB} shift about 19.3 V and the charge loss about 13.4% in such a sample under the ±15V gate voltage stress after 10^4 sec retention time test. Regardless of applied bias direction, the sample containing HfO$_2$/SiO$_2$ layer exhibited the leakage current density as low as 150 nA/cm^2 as revealed by the current-voltage (I-V) measurement. This effectively suppresses the electron injection between gate electrode and charge trapping layer and leads to a substantial enhancement of NFGM characteristics.

INTRODUCTION

Nonvolatile floating gate device (NFGM) comprised of metallic nanocrystals (NCs) as the discrete charge-storage traps has been widely investigated in recent years. The charge trapping layers comprised of transition metals such as cobalt (Co) [1], platinum (Pt) [2] and nickel (Ni) [3] NCs embedded in high-k dielectrics have attracted a lot of attentions due to the advantages including higher density of states around Fermi level, wide range of work functions, smaller energy disturbance and stronger coupling with the conduction channel [4]. Such NFGM devices exhibited better memory window (ΔV_{FB}) shifts and superior charge retention characteristics in comparison with those containing semiconductor NCs. Nevertheless, some drawbacks, for instance, the metal/dielectric oxides interaction and element interdiffusion during subsequent integration and high-temperature annealed processes would deteriorate the retention performance of NFGM [5]. Meanwhile, for improving the program/erase efficiency and charge retention property, HfO$_2$ has been recognized as one of the promising blocking oxides to replace the conventional SiO$_2$ owing to its relatively high dielectric constant, wide bandgap and large conduction band-offset with respect to Si [6]. However, HfO$_2$ is relatively poor in inhibiting the oxygen (O) diffusion and boron (B) penetration when post annealing temperature exceeds about 400°C. This leads to a high leakage current and low carrier mobility due to the onset of recrystallization [7]. In addition, a thin interfacial layer induced by oxidization and/or Hf diffusion was inevitably formed between HfO_2 and Si substrate during either the deposition or post-annealing process [8]. Presently, the nitrogen (N) or aluminum (Al) incorporations have been proposed to suppress the recrystallization and block the oxygen diffusion towards interface to preserve the integrity of HfO_2 layer. However, the doping methods might interrupt the stoichiometry of deposited layer and escalate the complexity of device fabrication processes [9]. In this work, nanocomposite thin film comprised of AgInSbTe (AIST) chalcogenide NCs embedded in SiO_2 matrix was adopted as the charge trapping layer due to its ultra fast phase-change rate and comparatively low recrystallization temperature [10]. Our previous study has demonstrated the NFGM device containing a sole AIST nanocomposite layer may have a good memory performance with ΔV_{FB} shift $\approx 6.9V$ at $\pm 8V$ gate voltage sweep [11]. In this work, HfO_2 or HfO_2/SiO_2 blocking oxide layer was incorporated in the devices and the enhancement of NFGM characteristics is presented as follows.

EXPERIMENTAL DETAILS

30 nm-thick AIST nanocomposite thin film was first deposited on Si substrate by using target-attachment method in a sputtering system with background pressure $\leq 2 \times 10^{-6}$ torr. During the deposition, the rf sputtering power = 100 W and the inlet Ar/N_2 gas flow ratio = 10:2 (in the unit of sccm). Afterward, a 7-nm thick HfO_2 thin film was deposited on nanocomposite layer to form the $HfO_2/AIST-SiO_2$ (HA) sample in the same sputtering system with rf sputtering power = 100W and inlet Ar gas flow = 10 sccm. As to $HfO_2/SiO_2/AIST-SiO_2$ (HSA) sample, a 7 nm-thick SiO_2 was first deposited on AIST nanocomposite layer by plasma-enhanced chemical vapor deposition (PECVD) at 250°C. Subsequently, a 7-nm thick HfO_2 layer was deposited on by using the process same as that for HA sample. Both HA and HSA sample were subjected to a 400°C/90-sec post annealing in atmospheric ambient to induce the recrystallization of AIST NCs. Finally, 300 nm-thick Al electrode layer was deposited on to complete the metal-insulator-semiconductor (MIS) device preparation. Microstructure and chemical status of elements in NFGM samples were characterized by a transmission electron microscopy (TEM, FEI TECNAI G2 F20 S-TWIN) equipped with an energy dispersive spectroscopy (EDX, Link ISIS 300) and an x-ray photoelectron spectroscopy (XPS, PHI Quantera SXM), respectively. Capacitance-voltage (C-V) and charge retention properties of NFGM devices were acquired to obtain the ΔV_{FB} shift at the frequency = 1 MHz by an HP 4284A precision LCR meter. The current-voltage (I-V) profiles were measured by an HP 4156B semiconductor parameter analyzer in conjunction with a probe tester (SANWA, WM-365A-1) in order to explore the conduction mechanisms across the device samples. All electrical measurements were performed in atmospheric ambient at room temperature.

RESULTS AND DISCUSSION

NFGM properties

Figure 1(a) presents the C-V profiles of HA sample under gate voltage sweep ranging from ±3V to ±15V. A counterclockwise hysteresis with significant ΔV_{FB} shift =11.6 V at ±15V voltage sweep can be observed, illustrating the saturated charge injection into the AIST NCs which act as the charge storage traps. The maximum storage charge density was found to be 1.0×10^{12} cm^{-2} according to the formula proposed by Maikap et $al.$ [12]. The asymmetrical loop shifts to negative bias side indicated the presence of positive fixed charges in nanocomposite layer, resulting in the defects such as oxygen vacancies in SiO$_2$ matrix. As to the HSA sample, it exhibits a significantly large ΔV_{FB} shift = 30.7 V and maximum charge density = 2.3×10^{13} cm^{-2} after ±23V voltage sweep as depicted in Fig. 1(b). Moreover, the ΔV_{FB} shift toward the positive bias side for HSA sample in comparison with the HA sample, implying the increasing charge storage in metallic NCs at positive bias.

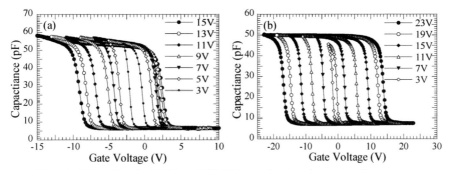

Figure1. C-V profiles for (a) HA and (b) HSA samples at various gate voltage sweep.

Figure 2(a) presents the charge retention characteristics of HA and HSA samples. The ΔV_{FB} shifts after electron (program) and hole (erase) injection $versus$ time were measured up to 10^4 sec. As shown in Fig. 2(a), HA sample presents a ΔV_{FB} shift = 7.22 V and the charge loss = 18.6% after applying the ±9V gate voltage stress for 10^4 sec. A dramatic enhancement of retention property is observed in HSA sample which exhibits a larger ΔV_{FB} shift = 19.32 V and the lower charge loss = 13.36% under the ±15V gate voltage stress for 10^4 sec. Figure 2(b) presents the profiles of leakage current density $versus$ applied bias field (J-E) of HA and HSA samples. Apparently, the leakage current of HSA sample is dramatically reduced in comparison with HA sample in particular in negative bias side. This is ascribed to the insertion of SiO$_2$ blocking oxide layer which improved the retention property by inhibiting the carriers injecting and charge tunneling between gate electrodes and charge trapping layer.

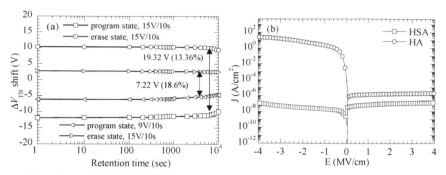

Figure 2. (a) Retention time characteristics and (b) *J-E* profiles of HA and HSA samples.

TEM characterization

Cross-sectional TEM (XTEM) images of HA and HSA samples are separately presented in Figs. 3(a) and 3(b). EDX analysis revealed the stoichiometric ratio of AIST is $Ag_{0.18}In_{0.33}Sb_{2.1}Te$, *i.e.*, AIST is the growth-dominated Sb_2Te phase doped with Ag and In elements. In both XTEM images, an about 3-nm thick SiO_x layer can be seen in between the region enriched with AIST NCs and Si substrate. This oxide layer might serve as the tunneling layer for charge transporting from Si substrate to Sb_2Te NC traps. On the contrary, there is a fuzzy interface in between HfO_2 and nanocomposite layers in HA sample whereas a SiO_2 layer visibly presents in the HSA sample. Notably, such a PECVD SiO_2 layer remains amorphous after the 400°C post annealing. It effectively inhibits the charge injection due to the absence of grain boundaries which may serve as the fast tunneling paths of charge carriers.

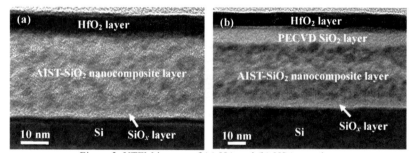

Figure 3. XTEM images of (a) HA and (b) HSA samples.

XPS analysis

Figures 4(a)-(c) depicts the XPS spectra and their de-convoluted profiles obtained by Gaussian curve fitting method [13] for the HA and HSA samples. Prior to the XPS analysis, an about 30-nm thick blocking oxides and part of nanocomposite layer were removed by *in-situ* Ar ion sputtering to expose the region enriched with AIST NCs. As shown by the Si $2p$ XPS profiles in Fig. 4(a), the thermal process of PECVD and post annealing promote the SiO_2 component and suppress the SiO_x component in AIST nanocomposite layer. This reduces the oxygen defects and is beneficial to the charge trapping on AIST NCs. The O $1s$ and Sb $4d$ XPS spectra shown in Figs. 4(b) and 4(c) indicate an increase of metallic Sb and Te phases in HSA sample, resulting in an improvement of charge storage capability and retention property in such a sample. The Hf $4d$ XPS depth profiles for HSA sample depicted in Fig. 4(d) reveals the HfO_2 seems to react with PECVD SiO_2 layer to form the $HfSiO_x$. Such a reaction impedes the Hf diffusion into the AIST nanocomposite layer and thus preserves the interface integrity, leading to the enhancement of NFGM characteristics.

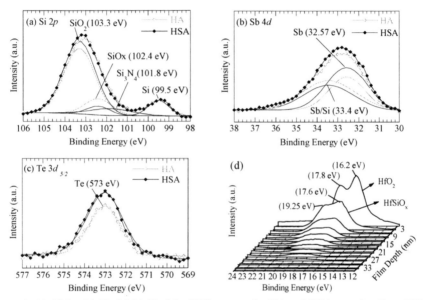

Figure 4. (a) Si $2p$ (b) Sb $4d$ (c) Te $3d_{5/2}$ XPS spectra for HA and HSA samples. Raw XPS profiles are decorated with solid or open diamonds while the corresponding curve fitting results are represented by solid and grey curves. (d) Hf $4d$ XPS depth profiles of HSA sample.

CONCLUSIONS

Performance enhancement of NFGM containing AIST nanocomposite as the charge storage traps by inserting an HfO_2 or HfO_2/SiO_2 blocking oxide layer is demonstrated in this study.

Significantly large ΔV_{FB} shift = 30.7 V and charge density = 2.3 $\times 10^{13}$ cm^{-2} at ± 23 V gate voltage sweep were achieved and the ΔV_{FB} shift about 19.4V and the charge loss about 13.4% were observed after the retention time test of 10^4 sec at ± 15V gate voltage stress in the NFGM sample containing HfO_2/SiO_2 layer. TEM and XPS characterizations indicated that inserted PECVD SiO_2 blocking layer is essential to the dramatic enhancement of NFGM characteristics since it not only impedes the Hf diffusion but also preserves the interface integrity in NFGM sample. In conjunction with the reduction of oxygen defects in the nanocomposite layer by the thermal process of PECVD and post annealing, a significant enhancement on NFGM characteristics can thus be achieved.

ACKNOWLEDGMENTS

The authors would like to thank to the support of National Science Council (NSC), Taiwan, R.O.C., under the contract No. NSC-97-2221-E-009-029-MY3. TEM and XPS analyses supported by Materials Analysis Technology Inc. and Instrument Center at National Tsing Hua University are also deeply acknowledged.

REFERENCES

1. B. Li, J. Liu, *J. Appl. Phys. Lett.* **101**, 124503 (2007).
2. S. K. Samanta, W. J. Yoo, G. Samudra, E. S. Tok, L. K. Bera, N. Balasubramanian, *Appl. Phys. Lett.* **87**, 113110 (2005).
3. Y.-S. Jang, J.-H. Yoon, *IEEE Trans. Electron Devices* **50**, 1823 (2003).
4. J. Kim, J. Yang, J. Lee, J. Hong, *Appl. Phys. Lett.* **92**, 013512 (2008).
5. Y. Pei, M. Nishijima, T. Fukushima, T. Tanaka, M. Koyanagi, *Appl. Phys. Lett.* **93**, 113115 (2008).
6. X. J. Wang, L. D. Zhang, M. Liu, J. P. Zhang, G. He, *Appl. Phys. Lett.* **92**, 122901 (2008).
7. M. Liu, Fang, Q.; G. He, L. Q. Zhu, L. D. Zhang, *Surf. Sci.* **252**, 8673 (2006).
8. S. J. Wang, J. W. Chai, Y. F. Dong, Y. P. Feng, N. Sutanto, J. S. Pan, A. C. H. Huan, *Appl. Phys. Lett.* **88**, 192103 (2006).
9. H. Wang, Y. Wang, J. Zhang, C. Ye, H. B. Wang, J. Feng, B. Y. Wang, Q. Li, Y. Jiang, *Appl. Phys. Lett.* **93**, 202904 (2008).
10. C. C. Chou, F. Y. Hung, T. S. Lui, *Scripta Materialia* **56**,1107 (2007).
11. K.-C. Chiang, T. -H. Hsieh, *IEEE Trans. Magn.* **47**, 656-662 (2010).
12. S. Maikap, S. Z. Rahaman, T. C. Tien, *Nanotechnology* **19**, 435202 (2008).
13. J. F. Moulder, W. F Stickle, P. E. Sobol, K. D. Bombem, *Handbook of X-ray Photoelectron Spectroscopy*, 2nd ed., Physical Electronics, Minnesota, 1992.

Resistive Switching Memories

Mater. Res. Soc. Symp. Proc. Vol. 1337 © 2011 Materials Research Society
DOI: 10.1557/opl.2011.856

Complementary Resistive Switches (CRS): High speed performance for the application in passive nanocrossbar arrays

Roland Rosezin[1], Eike Linn[2], Lutz Nielen[2], Carsten Kügeler[1], Rainer Bruchhaus[1], and Rainer Waser[1,2]

[1]Peter Grünberg Institut, Forschungszentrum Jülich GmbH, 52425 Jülich, Germany and JARA –

Fundamentals for Future Information Technology, Forschungszentrum Jülich GmbH, 52425

Jülich, Germany

[2]Institut für Werkstoffe der Elektrotechnik II, RWTH Aachen, 52074 Aachen, Germany

ABSTRACT

In this report, the fabrication and electrical characterization of fully vertically integrated complementary resistive switches (CRS), which consist of two anti-serially connected $Cu-SiO_2$ memristive elements, is presented. The resulting CRS cells are initialized by a simple procedure and show high uniformity of resistance states afterwards. Furthermore, the CRS cells show high switching speeds below 50 ns, making them excellent building blocks for next generation non-volatile memory based on passive nanocrossbar arrays.

INTRODUCTION

The term "resistive switching" denotes a vast field of voltage induced resistance change effects observed in simple, two terminal metal-isolator-metal structures [1]. The non-volatile nature of this resistance change as well as the high operation speed and scaling potential make devices based on this effect prime candidates for future high density data storage applications [2]. The field of resistance change phenomena can be divided into different classes according to the underlying physical principle. A number of material systems can be classified as electrochemical metallization (ECM) and valence change mechanism (VCM) type resistive switching materials [3]. Although the physical principles of the switching in these two classes are different, both show bipolar resistive switching, requiring different voltage polarities for switching to the low resistance state (LRS) and high resistance state (HRS) [4]. A very promising architecture for the integration of resistive switching devices is the passive crossbar array. In these crossbar arrays, a minimum area consumption of $4F^2$ (F being the feature size) can be achieved, because each storage node consists only of a resistive switching element. Due to the lack of select transistors, current sneak paths can occur, which strongly limit the practical size of these arrays. For bipolar resistive switching materials, a promising solution has been suggested to alleviate sneak paths and make large, passive crossbar arrays possible: the complementary resistive switch (CRS) [5]. In CRS cells, two bipolar switching devices are connected anti-serially. Thus, the information is stored as a combination of resistance states (HRS/LRS and LRS/HRS), which results in an overall high resistance for the CRS cell. Consequently, no pattern dependence can give rise to sneak paths. In this paper, we present the fabrication and electrical characterization of CRS cells based on $Cu-SiO_2$ memristive elements [6]. These elements were chosen due to their excellent switching performance and can be assigned to the ECM class of resistive switching materials [7].

EXPERIMENT

The fabrication of CRS cells begins with structuring the bottom electrode, in this case Pt, using optical lithography. Diameters ranging from 2 μm to 10 μm were achieved this way. On the Pt electrode, which is 30 nm thick, a 20 nm layer of SiO_2 is deposited by sputtering (175 W, 60 sccm Ar). For ECM type resistive switching devices, the active electrode is significant for the switching performance. Here, Cu is chosen. Thus, by depositing 20 nm Cu on top by evaporation, the first resistive switching element is completed. To increase the stability of this Cu layer, a thin Pt layer (15 nm) is deposited before evaporating another 20 nm of Cu.

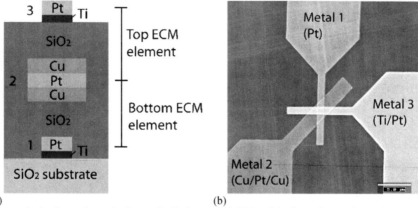

(a)   (b)

Figure 1. A schematic stack of a vertically integrated CRS cell is shown in panel (a). The constituting ECM elements and the electrodes are marked. Panel (b) shows a scanning electron micrograph of a CRS cell. The electrodes are marked accordingly.

The resulting Cu/Pt/Cu stack is structured by optical lithography and Ar etching. In a possible application, the Cu/Pt/Cu can be structured as a square metal patch, but here, for testing purposes, a probe electrode is defined. The advantages of this probe electrode are described in [8]. To complete the second memristive Cu-SiO_2 element and thus to complete the CRS cell, another 20 nm layer of SiO_2 is sputter deposited, and on top of a 5 nm Ti adhesion layer, a Pt top electrode is defined by optical lithography. The thickness of the Pt electrode was chosen to be 50 nm. A sketch of the final CRS cell stack is shown in Fig. 1(a), a corresponding SEM image is shown in Fig. 1(b).

To evaluate the electrical performance, quasi-static measurements were performed using an Agilent B1500A semiconductor analyzer. Short voltage pulses were applied using an Agilent 81110A pattern generator and the change in resistance was subsequently evaluated by quasi-static measurements.

DISCUSSION

The only resistance combination, which is not used during CRS operation, is the HRS/HRS state [5]. It is assumed that this state exists only directly after fabrication and does not restrict CRS operation. To investigate this, the initial resistance of top and bottom memristive element is determined quasi-statically using the probe electrode. A voltage amplitude of 100 mV is used which has no influence on this initial resistance. Fig. 2 shows the broad distribution of the resulting initial resistance for 20 CRS cells with a diameter of 2 μm.

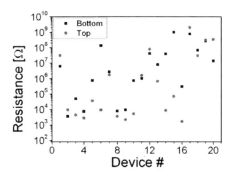

Figure 2. Initial resistance of top and bottom ECM element, as defined in Fig. 1 (a). Many different initial CRS resistance combinations are measured, among these also HRS/HRS combinations.

Several devices (e.g. 1, 7, 11, 12...) feature a HRS/HRS combination initially. Previous studies have shown that in general the forming procedure of a single ECM element requires only a voltage pulse of sufficient magnitude to assumingly build a preferential path into the matrix material. Thus, one can speculate that a sufficient voltage pulse enables to switch the corresponding element to the LRS although the total current through the cell will be limited by the other element, which is in the HRS. The resulting CRS state would then be LRS/HRS from which normal operation can resume. To verify this speculation, 5 V quasi-static staircase sweeps were applied to the bottom electrode (Fig. 1(a), #1) of the CRS cell, while setting the top electrode (Fig. 1(a), #3) to ground potential and the probe electrode (Fig. 1(a), #2) to floating potential. Afterwards, the resistance of the respective elements was measured by means of the probe electrode, resulting in the resistance distribution shown in Fig. 3 (a). Clearly, all devices exhibit nearly the same LRS resistances with similar HRS resistances, only one of the devices tested ceased to function. Thus, the CRS devices can be initialized by a simple write voltage pulse, e.g. during wafer level testing. To further test the viability of this procedure, a write pulse with opposite polarity was applied. The analysis of the results, which are shown in Fig. 3 (b), reveals that the CRS devices function as desired, underlining the feasibility of the suggested principle of initialization.

Apart from an easy forming procedure, fast switching during operation is required for a possible application as non-volatile memory device. To test the fast switching behavior, CRS cells were switched into a defined state and their resistance was checked.

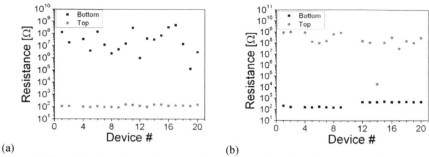

(a)   (b)

Figure 3. Panel (a) shows the resistances as determined using the probe electrode after the application of a voltage sweep of 5 V amplitude to the bottom electrode. Panel (b) shows the resistances after a subsequent write operation into the HRS/LRS state by applying 5 V to the top electrode.

Afterwards, a voltage pulse of 10 V magnitude and 50 ns width was applied to the top electrode in order to switch the device. Then, the success of the switching was determined by quasi-static measurements with an amplitude of 100 mV, before a pulse of the same polarity applied to the bottom electrode is used to switch the device back into the original state (Fig. 4). Thus, the experiments prove the fast switching speed of such CRS cells.

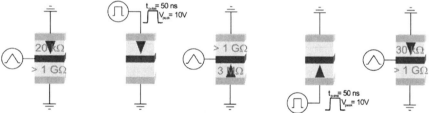

Figure 4. Measurement procedure for the fast pulse switching of CRS cells. The initial resistances of each element were checked using a quasi-static sweep with small voltages, indicated by the triangular waveform. Next, a pulse of 50 ns width and 10 V amplitude is applied to the device. The resulting resistances are again determined by a staircase sweep. To switch the device to the other state, the pulse is applied to the bottom electrode.

CONCLUSIONS

We report on the fabrication, initialization and high-speed operation of fully vertically integrated CRS cells. With Cu-SiO₂ memristive cells, elements based on a very promising ECM type resistive switching material system were chosen. Fully vertically integrated cells were fabricated and tested regarding their initial resistance. It was found that the devices exhibit a large range of possible initial resistances and combinations but it could be proven that a write voltage pulse suffices to initialize these cells into a defined CRS state. Then, normal CRS operation can be started. The switching speed of these cells was tested by applying voltage pulses of 10 V magnitude and 50 ns pulse width. By determining the resistance of the respective elements using quasi-static sweeps, the successful switching using these pulses was unambiguously proven. Further work includes tailoring of device resistances and switching voltages so that low power and low voltage operations become possible. In conclusion, CRS based crossbar arrays are an exciting option for non-volatile memory devices and offer the prospect of a combined logic and memory functionality [9].

ACKNOWLEDGMENTS

The authors would like to thank R. Borowski and H. Haselier for assistance during sample preparation and F. Lentz and C. Hermes for fruitful discussions.

REFERENCES

1. R. Waser and M. Aono, *Nat. Mater.* **6**, 833 (2007).
2. D. B. Strukov and R. S. Williams, *Appl. Phys. A-Mater. Sci. Process.* **94**, 515 (2009).
3. R. Waser, R. Dittmann, G. Staikov, and K. Szot, *Adv. Mater.* **21**, 2632 (2009).
4. C. Kügeler, R. Rosezin, E. Linn, R. Bruchhaus, and R. Waser, *Appl. Phys. A - Mater. Sci. Process.* **102**, 791 (2011).
5. E. Linn, R. Rosezin, C. Kügeler, and R. Waser, *Nat. Mater.* **9**, 403 (2010).
6. D. B. Strukov, G. S. Snider, D. R. Stewart, and R. S. Williams, *Nature* **453**, 80 (2008).
7. C. Schindler, M. Weides, M. N. Kozicki, and R. Waser, *Appl. Phys. Lett.* **92**, 122910 (2008).
8. R. Rosezin, E. Linn, L. Nielen, C. Kügeler, R. Bruchhaus, and R. Waser, *IEEE Electron Device Letters* **32**, 191 (2011).
9. R. Rosezin, E. Linn, C. Kügeler, R. Bruchhaus, and R. Waser, *IEEE Electron Device Letters,* in press, (2011).

Mater. Res. Soc. Symp. Proc. Vol. 1337 © 2011 Materials Research Society
DOI: 10.1557/opl.2011.857

Influence of Copper on the Switching Properties of Hafnium Oxide-Based Resistive Memory

B.D. Briggs[1], S.M. Bishop[1], K.D. Leedy[2], B. Butcher[1], R. L. Moore[1], S. W. Novak[1], and N.C. Cady[1]

[1]University at Albany, SUNY, College of Nanoscale Science and Engineering, Albany, NY 12203, U.S.A.
[2]Air Force Research Laboratory, 2241 Avionics Circle, Dayton, OH, 45433, U.S.A.

ABSTRACT

Hafnium oxide-based resistive memory devices have been fabricated on copper bottom electrodes. The HfO_x active layers in these devices were deposited by atomic layer deposition at 250 °C with tetrakis(dimethylamido)hafnium(IV) as the metal precursor and an O_2 plasma as the reactant. Depth profiles of the HfO_x by x-ray photoelectron spectroscopy and secondary ion mass spectroscopy revealed a copper concentration on the order of five atomic percent throughout the HfO_x film. This phenomenon has not been previously reported in resistive switching literature and therefore may have gone unnoticed by other investigators. The MIM structures fabricated from the HfOx exhibited non-polar behavior, independent of the top metal electrode (Ni, Pt, Al, Au). These results are analogous to the non-polar switching behavior observed by Yang *et al.* [2] for intentionally Cu-doped HfOx resistive memory devices. The distinguishing characteristic of the material structure produced in this research is that the copper concentration increases to 60 % in a conducting surface copper oxide layer ~20 nm thick. Lastly, the results from both sweep- and pulse-mode current-voltage measurements are presented and preliminary work on fabricating sub-100 nm devices is summarized.

INTRODUCTION

Transition metal oxide resistive memory devices are a leading candidate for next generation non-volatile storage. When compared to current CMOS NAND flash, resistive memory has the potential to offer lower power operation, increased density through a simpler fabrication process and ultimately, faster operation. Hafnium oxide has been selected as the transition metal oxide for this work due to its large band gap (5.8 eV), the resulting low off-state leakage current, and its wide spread use in CMOS manufacturing as a high-k gate dielectric. Hafnium oxide has been previously studied as the active layer in resistive memory devices by multiple groups. These investigations have demonstrated devices with stable, long-term read/write endurance [1-3], low switching energy [4], and high on/off ratios [2]. Previous work has shown the effects of Cu on the performance of HfO_x based resistive memory, by the insertion of a Cu layer between the HfO_x used to intentionally dope the HfO_x [2] or using Cu as a top electrode as a reservoir of metallic ions, thought to form conductive filaments during switching [3]. Of particular interest to our work is the role of copper doping in HfO_2 films created by atomic layer deposition on copper.

EXPERIMENTAL PROCEDURE

A starting substrate of SiO_2/SiN was fabricated on 300mm silicon wafers using standard chemical vapor deposition techniques. Atop the substrate, Cu/Ta/TaN were deposited by physical vapor deposition (PVD), acting as the electroplating seed, adhesion layer, and diffusion barrier, respectively. To create the bottom electrode, 1μm of electroplated (ECD) Cu was deposited on the Cu seed layer. Chemical mechanical planarization (CMP) was then used to level and polish the ECD Cu. The deposition of HfO_x layer was carried out by atomic layer deposition (ALD), with a chuck temperature of 250 °C and a chamber pressure of 0.19 torr. Tetrakis (dimethylamido)hafnium(IV) was used as the metal-organic precursor and a 300 W RF O_2 plasma as the reactant. The target thickness of HfO_x was 50 nm, which required 603 ALD cycles and 6.23 hrs of deposition time.

Microscale devices were fabricated by patterning top electrodes using either a shadow mask or a conventional photolithography-based lift-off process. The top electrodes (Pt, Ni, Al, Au) were deposited with thicknesses of 75-100 nm by electron beam evaporation; the resulting contacts ranged in diameter from 25-100 μm. Nanoscale devices on the order of 70 nm were fabricated using a damascene copper via process. Vias were etched into the SiO_2 layer on the same substrate detailed above (Cu/Ta/TaN/SiO_2/SiN/Si) using optical lithography and a reactive ion etch process. Cu/Ta/TaN was deposited by PVD into the vias, to act as a diffusion barrier and a seed for electroplating. Copper was then electroplated into the vias and CMP was used to remove the Cu overburden and planarize the wafer surface. The same HfO_x deposition conditions and top contact process detailed above were used to finish the device fabrication.

Xray photoelectron spectroscopy (XPS) surface characterization and sputter depth profiling of the as-deposited HfO_x structure was performed using a ThermoFisher Thetaprobe equipped with a hemispherical analyzer and a monochromated Al Kα x-ray source (1486.6 eV) operated at a 100 W/400 μm spot mode for area-averaged analyses. Sputter depth profiling was accomplished using an argon ion gun operated at 2.5 keV with a sample current of ~1.4 μA, yielding a sputter rate of 4 nm/minute for SiO2. In addition, a completed device structure was depth profiled with secondary ion mass spectroscopy (SIMS). Depth profiles were acquired using an IonTof V time-of-flight SIMS system. The sputtering beam was 2keV Cs at 45 degrees incidence and the pulsed analysis beam was 25keV Bi+ at 45 degrees. Negative secondary ions were collected from a 30x30 micron area. The primary beam pulse time yielded a mass resolution of 7000-8000, allowing full separation of interfering masses.

Switching characteristics were investigated using sweep based current voltage (I-V) electrical measurements on an Agilent B1500A semiconductor parameter analyzer. Preliminary pulse measurements were performed using a Keithley 4200 semiconductor parameter analyzer and Agilent 81110A pulse generator. For pulse measurements devices were connected in a 1T1R configuration (ReRAM attached to the drain) using an external transistor. This was done to allow for current compliance capability during pulsing.

RESULTS & DISCUSSION

The XPS depth profile of an as-deposited HfO_x film is shown in Figure 1A. There are three noteworthy points from these data. First, the profile shows that the HfO_x is not stoichiometric because the [O]/[Hf] ratio is equal to 1.1. Second, there is a significant (unintentional) copper impurity concentration present in the HfO_x. The latter is on the order of five atomic percent throughout the portion of the HfO_x layer that was analyzed. Even more surprising is the increase in the copper concentration at the beginning of the profile. The copper concentration in this region varies from a maximum of 60%, to 32% at the sample surface. Concurrently, the oxygen profile varies before stabilizing at 50% in the HfO_x layer. These data clearly indicate the final point; a copper oxide layer of varying stoichiometry resides on the surface of the HfO_x. The presence of this Cu_xO layer has not been reported previously. In addition, this layer could play a significant role in the electrical properties of these devices because Cu_xO has also been shown to be a resistive memory material [5].

Because it has better depth resolution and sensitivity to trace concentrations than XPS, SIMS was also employed to characterize the copper within the HfO_x. Figure 1B shows the results from a SIMS analysis of a $Pt/HfO_x/Cu$ device structure. This device was profiled "as fabricated" and was therefore not influenced by any electrical bias before analysis. At the uppermost interface (Pt/HfO_x) the Cu signal peaks, along with O, indicating the presence of the Cu_xO layer. The estimated thickness of the Cu_xO is less than 20 nm. The accuracy of this thickness measurement from SIMS is limited by the differential sputtering rates of the materials in the MIM device structure. After the peak concentration of Cu at the interface, the concentration declines continuously, until another spike in Cu is observed at the beginning of the bulk Cu electrode. This is not an expected result. Normally, in a bulk diffusion process, we would expect a decreasing concentration of Cu from the bulk bottom electrode to the surface, the opposite of what is observed. Further investigation is therefore needed to understand the profile of Cu observed in these structures as well as the trace concentration of Pt observed throughout the HfO_x region. Subsequent X-ray diffraction studies (not shown) indicated that the bulk structure of the film is amorphous, although this does not preclude regions of nanocrystallinity.

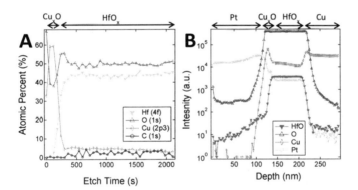

Figure 1. (A) XPS depth profile of ALD HfO_x on Cu. The uppermost portion of the profile shows Cu_xO layer formation. (B) SIMS depth profile of a $Pt/HfO_x/Cu$ MIM device, confirming the presence of an interfacial Cu_xO layer.

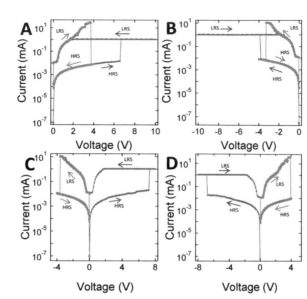

Figure 2. Current voltage measurements of a Pt/HfO$_x$/Cu MIM. (A-B) Unipolar operation, independent of bias direction. (C-D) Bipolar operation, with set voltage in both positive and negative polarity.

Top electrode biased current voltage measurements show both unipolar Figure 2A-B and bipolar Figure 2C-D resistive switching behavior. This behavior is known as "non-polar" resistive switching, in which only the magnitude of the voltage, not the polarity, dictates switching. The device in Figure 2. exhibited a change in resistance state during the application of a voltage sweep $[0 \rightarrow V_{set}]$ with $V_{set} = |6\text{-}8|$ V. (Note: all sets used a 1 mA current compliance which was found to be the optimal value for these devices). Transitions from a LRS to HRS were where observed with a follow-up voltage sweep $[0 \rightarrow V_{reset}]$ of either polarity, with no current compliance. V_{reset} was consistent during each cycle at around $|4|$ V. Previous studies report that HfO$_x$ can behave as a unipolar [4], bipolar [3] or non-polar [1-2] switch . To the best of the author's knowledge, this is the first report of ALD HfO$_x$ on Cu for resistive memory. Our process shows a significant, 5% Cu doping, very similar to the work of Y. Wang *et al.*[2], who used a similar Cu doped HfO$_x$ MIM (Cu/HfO$_x$/Cu/HfO$_x$/Pt) structure where a thin Cu layer was deposited between the HfO$_x$ to dope it, in an attempt to improve switching performance. This group attributed the non-polar switching phenomena to the formation of Cu rich filaments, which form a low resistance conduction pathway. Our results of non-polar switching behavior agree very closely with the work of Y. Wang *et al.* [2]. The unintended formation of the Cu$_x$O layer by ALD, does not appear to influence on the root electronic mechanism, whether it be filamentary in nature, or an undiscovered phenomenon causing the non-polar resistive switching in HfO$_x$. Devices exhibited sweep mode endurance of around 2-15 cycles before failure (as shorts). This is likely due to the large quantity of power applied to the device during sweep mode measurements.

To alleviate this stress we employed pulse based measurements for the set operation. Figure 3A shows the reset sweep measurements for multiple cycles after a pulsed set (pulse width of 10 μs with a pulse height of 3 V). There is no current measured for the first -0.7 V of the reset curve, due to the bias scheme of the device in the 1T1R configuration (Figure 3B). In this configuration, a critical field is needed to overcome the drain to well junction barrier. Pulse based sets greatly improved the endurance of the devices (>30 cycles), and future work will include a complete set and reset pulsing study.

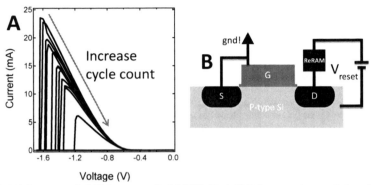

Figure 3. (A) Sweep mode I-V for reset, of a Ni/HfO$_x$/Cu MIM. Set operation carried out by application of pulse. (B) 1T1R configuration for reset I-V sweep.

Characterization of the nanoscale, via-based devices is shown in the focused ion beam milled, scanning electron microscopy (FIB-SEM) micrograph in Figure 4A. This image depicts a 200 nm Cu via array on a blanket Cu bottom electrode with 30 nm of ALD HfO$_x$. The depletion of Cu at the surface of the vias is likely occurring due to the same phenomenon which caused formation of Cu$_x$O on the HfO$_x$ surface, as described above. Future work will determine if Cu$_x$O is also present at the surface, post ALD of HfO$_x$.

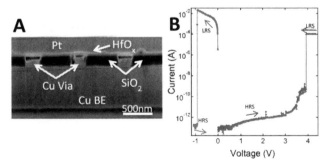

Figure 4. (A) FIB-SEM micrograph of 200nm Cu vias post ALD of HfO$_x$. Note Cu depletion at the surface of the Cu-filled vias. (B) Bipolar I-V of a 70nm Ni/HfO$_x$/Cu via device.

Sweep mode electrical measurements of the via based devices were performed in the same setup as the bulk film devices. Log plots of the I-V curves for a 70nm via are shown in Figure 4B, indicating bipolar behavior. V_{set} was below 5 V with a 100 μA current compliance (used to limit the total current density through the 70nm via). V_{reset} was less than 1 V, which yielded high I_{on}/I_{off} ratios of 10^{10} for these nanoscale devices.

CONCLUSION

ReRAM devices using ALD HfO_x were fabricated on Cu bottom electrodes, which resulted in the unintentional doping of the HfO_x and a stoichiometry of $x \approx 1.1$. XPS and SIMS depth profiles showed that a ~20nm thick Cu_xO layer formed at the surface of the HfO_x. I-V data show non-polar resistive switching properties independent of top electrode material. Comparison with previous work shows a good fit between Cu-doped HfO_x MIM behavior. Use of pulse mode measurement in a 1T1R configuration greatly improved endurance, by reducing total power dissipation stress during the set operation. Preliminary work on nanoscale via-based devices show bipolar switching characteristics with high I_{on}/I_{off} of 10^{10}. SEM images indicate, however, a large amount of Cu depletion at the surface of the via, due to the ALD of HfO_x. Future work will emphasize the influence of the unintentional Cu_xO layers role in resistive switching of HfO_x. Challenges remain, but HfO_x on Cu is a BEOL CMOS fabrication compatible non-volatile memory candidate with extremely promising electrical performance.

ACKNOWLEDGMENTS

This research was sponsored by the Air Force Research Laboratory awards FA87500910231 and FA87501110008. The authors would like to acknowledge the Center for Semiconductor Research at the University at Albany-SUNY for wafer development and processing, Dr. Joseph Van Nostrand, AFRL-RI, for programmatic and scientific support, and the Rensselaer Polytechnic Institute's Micro and Nano Fabrication Clean Room.

REFERENCES

1. V. Jousseaume, et al. Solid-State Electron. Vol. **58**, Issue 1, 62-67 (2011)
2. Y. Wang, et al. Nanotechnology. **21**, 045202 (2010)
3. M. Haemori, T. Nagata, and T. Chikyow. Applied. Physics Express. **2**, 061401 (2010)
4. Ch. Walcyzk, et al. App. Phys. Lett. **105**, 114103 (2009)
5. H. Lv, et al. App. Phys. Lett. **94**, 213502 (2009)

Mater. Res. Soc. Symp. Proc. Vol. 1337 © 2011 Materials Research Society
DOI: 10.1557/opl.2011.982

Fabrication and Characterization of Copper Oxide Resistive Memory Devices

S.M. Bishop[1], B.D. Briggs[1], Z.P. Rice[1], S. Addepalli[1], N.R. McDonald[2], and N.C. Cady[1]
[1]University at Albany, SUNY, College of Nanoscale Science and Engineering, Albany, NY
12203, U.S.A
[2]Air Force Research Laboratory/RITC, Rome, NY 13441, USA

ABSTRACT

Three synthesis techniques have been explored as routes to produce copper oxide for use in resistive memory devices (RMDs). The major results and their impact on device current-voltage characteristics are summarized. The majority of the devices fabricated from thermally oxidized copper exhibited a diode-like behavior independent of the top electrode. When these devices were etched to form mesa structures, bipolar switching was observed with set voltages <2.5 V, reset voltages <(-1) V and R_{OFF}/R_{ON} ~10^3-10^4. Bipolar switching behavior was also observed for devices fabricated from copper oxide synthesized by RT plasma oxidation (R_{OFF}/R_{ON} up to 10^8). Voiding at the copper-copper oxide interface occurred in films produced by thermal and plasma oxidation performed at ≥200°C. The copper oxide synthesized by reactive sputtering had large areas of open volume in the microstructure; this resulted in short circuited devices because of electrical contact between the bottom and top electrodes. The results for fabricating copper oxide into ≤100 nm features are also discussed.

INTRODUCTION

Resistive memory devices (RMDs) have the potential to revolutionize computing beyond today's transistor-based systems. Copper oxide is one of the transition metal oxides that exhibits resistance-based memory. Copper oxide RMDs have been produced with R_{OFF}/R_{ON} ratios up to 10^5 [1] and endurance values of >10^4 write/erase cycles [1]. Data retention times of >10 years have also been predicted [2,3]. Despite this progress, there is a significant lack of information available in the open literature on the fabrication of these devices. In this work, a systematic investigation has been performed to determine how the oxide synthesis technique, film properties, and the device fabrication process impact the switching behavior of RMDs. The technical challenges encountered with each synthesis technique will be discussed, and the process conditions employed to produce switchable oxide thin films will be introduced.

EXPERIMENTAL PROCEDURE

Copper oxide was synthesized by oxidation and reactive sputtering on top of 1 μm thick copper thin films that were electroplated on Ta/TaN/SiO$_x$/Si$_x$N$_y$/Si. Thermal oxidation of the copper electrode was performed at 200-400°C in air at atmospheric pressure. Plasma oxidation was performed at RT and 280°C with an oxygen flow rate of 1-14 slm, a pressure range of 0.5-2.5 Torr, and an RF power range of 300-1000 Watts. Copper oxide was deposited by room temperature reactive sputtering at 1×10^{-2} Torr and 2000 W; the oxygen flow rate was maintained at 20 sccm, while the argon flow rate ranged from 3.5-20 sccm. Scanning electron microscopy (SEM), atomic force microscopy (AFM), and transmission electron microscopy (TEM) were employed to analyze the microstructure of the resulting films. X-ray diffraction (XRD) was utilized to determine the phase(s) of copper oxide present. Devices were fabricated by patterning top electrodes using either a shadow mask or a conventional photolithography-based lift-off process. Aluminum, nickel or platinum were deposited with thicknesses of 75-100 nm by electron beam evaporation; the resulting contacts ranged in diameter from 25-1000 μm. Mesa devices were fabricated for select devices. Using the top electrode as an etch mask, concentrated phosphoric acid (85 vol. %) was used to remove the copper oxide layer between top electrodes. The switching properties of the devices were analyzed using sweep mode current-voltage (IV) measurements.

To understand how the above synthesis techniques scale down to nanometer scale features, two substrates were employed that contained vias patterned into silicon oxide. Substrate (1) contained empty vias and substrate (2) had copper filled vias. For both substrates, the patterned silicon oxide resided above a 1 micron thick copper layer that was electroplated onto Ta/TaN/SiO$_x$/Si$_x$N$_y$/Si. On-die the via diameters ranged from 33 nm to 10 microns. Copper oxide was deposited on substrate (1) and thermal oxidation was used to convert the surface of the copper filled vias in substrate (2) to copper oxide; the deposition and oxidation conditions were the same as those outlined above. Because the deposited copper oxide overfilled the patterned regions, a chemical-mechanical planarization (CMP) step was needed to remove the overburden.

RESULTS AND DISCUSSION

Copper oxide thin films produced by thermal oxidation over the temperature range 200-400°C exhibited rough surfaces. Figure 1(a) shows the AFM height image of a copper oxide films synthesized at 300°C for 60 min. The RMS (root-mean-square) roughness of the image is 32 nm. The topography of the surface is evident from the 100 nm height scale of the image. Cuprous Oxide (Cu$_2$O) was the primary phase for all of the films produced by thermal oxidation; their XRD patterns were similar to Fig. 4(a) below. The majority of the devices fabricated from thermally oxidized copper exhibited a diode-like behavior as shown in Fig. 1(b). The diode-like behavior was observed in as-fabricated devices and after a large increase in current, similar to the set process common to RMDs. The diode behavior did not change with time or bias conditions and thus multiple resistance states were not observed for these devices. The diode-like IV behavior occurred independent of the top electrode used. Some devices fabricated from thermally oxidized copper with large area contacts (1 mm in diameter) exhibited bipolar

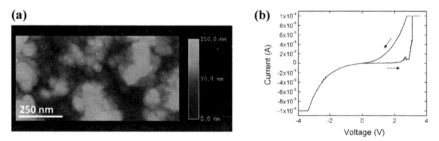

(a)

(b)

Figure 1. (a) AFM height image of the surface of thermally oxidized copper (300°C and 60 min.) and (b) the diode-like current-voltage behavior from a device fabricated from the same sample with an Al top electrode (100 nm).

switching; however, the resistance ratio (R_{OFF}/R_{ON}) was low (~1.5). The latter is due presumably to the large area of the contacts giving rise to low off-state resistance values.

Plasma oxidation of copper was also investigated as a route to synthesize copper oxide. Our results show that temperature plays a significant role in the oxidation rate and the microstructure in the resulting film. Plasma oxidation at room temperature yielded continuous thin films ranging in thickness from 5-10 nm, independent of the other process conditions. Figure 2(a) shows a TEM image of a ~9 nm layer of copper oxide. Bipolar switching behavior was observed for devices fabricated from copper oxide synthesized by room temperature plasma oxidation. Figure 2(b) shows the low resistance (LRS) and high resistance states (HRS) of a Al/Cu$_x$O/Cu device. The set voltage and the reset voltage were 2.8 V and -0.3 V, respectively, and the $R_{OFF}/R_{ON} = 10^8$.

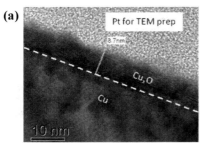
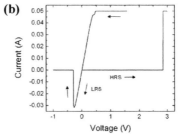

Figure 2. (a) TEM cross-sectional image showing a continuous thin film of RT plasma oxidized copper (300 Watts, 14 slm, 0.5 Torr, and 20 min.). (b) The bipolar switching behavior from a device fabricated from the same wafer with an Al top electrode (100 nm).

The copper oxide films produced by plasma oxidation at 280°C were significantly thicker than those created at room temperature. Figure 3(a) shows a TEM image of a copper oxide film (>700 nm thick) that was created by high temperature plasma oxidation. Although it is not immediately apparent from Fig. 3(a), the thickness of these films varied dramatically (>10% thickness uniformity was observed), resulting in significant surface roughness/topography. Devices fabricated from the high temperature plasma oxide exhibited the same diode behavior described above and shown in Fig. 1(b).

Voiding at the copper-copper oxide interface was observed in films produced by thermal and plasma oxidation at temperatures ≥200°C. These defects are immediately apparent in Fig. 3(a); voids up to 200 nm in width separate the copper and the plasma oxide. Figure 3(b) also shows a high degree of interfacial voiding in the cross section of a thermal oxide. Interfacial voids are clearly a problem for RMDs because they influence carrier transport between copper and copper oxide and thereby increase the resistance of the device. It is a noteworthy observation that the oxides that exhibit interfacial voiding also show the diode IV behavior. The latter is still being investigated. Other researchers have shown these defects are a common problem associated with copper oxidation and interfacial voids can form, for example, in areas with high impurity concentrations [4]. We are also investigating the origin of interfacial voiding in our process because these defects become more significant when scaling devices to smaller dimensions (see below).

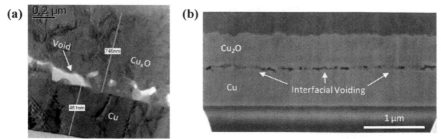

Figure 3. Voiding at the interface between copper and copper oxide synthesized by (a) plasma oxidation at 280°C (1000 Watts, 1 slm, 2.5 Torr, and 30 min.) and (b) thermal oxidation (300°C and 60 min.)

The copper oxide deposited by reactive sputtering had a polycrystalline microstructure with Cu_2O being the dominant phase present in the thin films. A representative XRD pattern for these films is shown in Fig. 4(a). The distinguishing feature of copper oxide synthesized by reactive sputtering over the oxides synthesized by oxidation was the large areas of open volume in the microstructure. As shown in Fig. 4(b), these areas were sufficiently large to permit electrical contact between the top and bottom electrodes. The latter lead to short circuiting (inset in Fig. 4(b)) in the devices fabricated from these films.

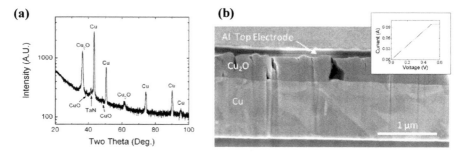

Figure 4. (a) XRD pattern showing Cu_2O is the primary phase in a copper oxide thin film deposited by reactive sputtering (Ar flow = 6.7 sccm). (b) SEM image showing the open volume in the copper oxide microstructure. This open volume allowed the top and bottom electrodes to make electrical contact, resulting in shorted devices (IV data inset).

A mesa etch process was developed to create isolated devices and prevent parasitic effects between individual devices. Of the chemistries suggested in ref. [5], phosphoric acid gave the highest degree of selectivity, where the copper oxide layer between top electrodes was removed and the amount of undercutting below the top electrode was minimized. Figure 5(a) shows a 100 μm device from the top and side (inset). The current-voltage characteristics of devices consisting of a continuous thermal copper oxide layer were compared with those that had been fabricated into individually defined mesa devices. Our preliminary results suggest that

58

individually-defined mesa devices exhibit more reliable resistive switching characteristics. Figure 5(b) shows the switching behavior for a thermal oxide mesa device. These data are markedly better than the diode behavior and the low resistance bipolar switching data discussed above for the continuous thermal oxide devices. Bipolar resistive switching was observed for mesa devices with both aluminum and platinum top electrodes. Nickel top electrodes were not utilized in the mesa device studies. Mesa devices with aluminum as the top electrode exhibited lower set voltages (<2.5 V) and higher R_{OFF}/R_{ON} ratios (10^3-10^4). The reset voltages were less than -1V for both Al and Pt top electrodes. We speculate that the improved switching properties of the mesa devices in this research are related to the increased current confinement in these devices over those with a continuous oxide layer. To improve the selectivity of the etch process and minimize the roughness shown in the inset in Fig. 5(a), a dry etch process is being developed similar to the work of Takano et al. [6].

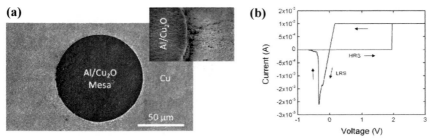

Figure 5. (a) Planview SEM image of a 100 μm (diameter) mesa device. The side profile is shown in the inset. The bipolar switching behavior for a mesa device on the same wafer is shown in (b); the top electrode was Al (100 nm).

There is a significant lack of information available in the open literature on the fabrication of nanoscale RMDs. The results of our preliminary work on synthesizing copper oxide into nanometer scale features is summarized here. Thermal oxidation of copper-filled vias created copper oxide in localized areas down to 33 nm in diameter; however the copper oxide separated from the underlying copper. Figure 6(a) shows a cross-sectional SEM image of ~250 nm vias that have been oxidized at 300°C for 20 min. The gap between the copper oxide and the copper remaining in the via is likely the result of copper being depleted from this region during oxidation and/or delamination induced by the lattice and thermal expansion mismatch between the materials. Copper oxide was also deposited into via structures by reactive sputtering. A representative SEM image of the wafer surface is shown in Fig. 6(b). The incomplete fill is clear. It is unclear if the large cracks in the surface of the copper oxide occurred during deposition or are a result of the post-deposition CMP step.

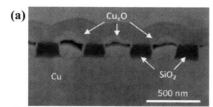 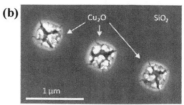

Figure 6. SEM images showing: (a) the voiding between copper oxide and the copper remaining in 250 nm vias after thermal oxidation (300°C and 20 min.) and (b) 500 nm vias (post-CMP) that are partially filled with sputtered deposited (Ar = 20 sccm) copper oxide. The layers above the Cu_xO in (a) are protective Pt films from the FIB sample preparation.

SUMMARY AND CONCLUSIONS

Copper oxide has been synthesized by thermal and plasma oxidation and reactive sputtering. Devices fabricated 1) from thin copper oxide layers and 2) into confined areas yielded the most robust switching characteristics. High oxidation rates produced copper oxide films with a defective microstructure. Voids at the copper-copper oxide interface made devices based on these films unsuitable for resistive memory applications. Further, interfacial voiding becomes much more significant as the device size decreases. This study has illustrated some of the challenges that will be encountered as copper oxide RMDs are scaled to smaller dimensions and integrated with CMOS.

ACKNOWLEDGMENTS

This research was sponsored by the Air Force Research Laboratory awards FA87500910231 and FA87501110008. The authors would like to acknowledge the Center for Semiconductor Research at the University at Albany-SUNY for wafer development and TEM metrology. SMB would also like to acknowledge T. Murray (Univ. at Albany) and J. Drumheller (Cornell Nanofabrication Facility) for assisting with this research and Dr. Joseph Van Nostrand, AFRL-RI for programmatic and scientific support.

REFERENCES

1. H.B. Lv, M. Yin, X.F. Fu, Y.L. Song, L. Tang, P. Zhou, C.H. Zhao, T.A. Tang, B.A. Chen, and Y.Y. Lin, IEEE. Elect. Dev. Lett. 29, 47 (2008).
2. R. Dong, D.S. Lee, W.F. Xiang, S.J. Oh, D.J. Seong, S.H. Heo, H.J. Choi, M.J. Kwon, S.N. Seo, M.B. Pyun, M. Hassan, and H. Hwang, Appl. Phys. Lett. 90, 042107 (2007).
3. A. Chen, S. Haddad, Y.C. Wu, T.N. Fang, S. Kaza, and Z. Lan, Appl. Phys. Lett. 92, 013503 (2008).
4. Y. Zhu, K. Mimura, and M. Isshiki, Oxid. Met. 61, 293 (2004).
5. CRC Handbook of Metal Etchants, edited by P. Walker and W.H. Tarn (CRC Press LLC, 1991), p. 348.
6. F. Takano, H. Shima, H. Muramatsu, Y. Kokaze, Y. Nishioka, K. Suu, H. Kishi, N.B. Arboleda, Jr., M. David, T. Roman, H. Kasai, and H. Arkinaga, Jap. J. Appl. Phys. 47, 6931 (2008).

Mater. Res. Soc. Symp. Proc. Vol. 1337 © 2011 Materials Research Society
DOI: 10.1557/opl.2011.1203

Influence of Process Parameters on Resistive Switching in MOCVD NiO Films

X.P. Wang[1], D.J. Wouters[1,3], M. Toeller[2], J. Meersschaut[1], L. Goux[1], Y.Y. Chen[3],
B. Govoreanu[1], L. Pantisano[1], R. Degraeve[1], M. Jurczak[1], L. Altimime[1], J. Kittl[1]

[1]IMEC, Kapeldreef 74, B-3001 Leuven, Belgium
[2]Tokyo Electron Limited, Akasaka Biz Tower, 3-1 Akasaka 5-chome, Minato-ku, Tokyo 107-6325 Japan
[3]Department of Electrical Engineering (ESAT), Katholieke Universiteit Leuven, B-3001 Leuven, Belgium

ABSTRACT

The unipolar resisitive switching properties of MOCVD deposited NiO in Ni/NiO/TiN stacks is reported. The switching quality is defined as function of RESET current and Roff/Ron ratio, and the importance of the Forming current and voltage on these parameters is discussed. The effect of structural stack variations as NiO thickness, Ti doping, and TiN thickness on the switching behavior of NiO is explained by the effect on the forming current and voltage conditions, and on Joule heating dissipation. Thinner NiO films, Ti doping, as well as thicker top electrode improve the switching quality by decreasing the RESET current and increasing the Roff/Ron ratio.

INTRODUCTION

NiO has become one of the prototype metal-oxide resistive switching materials since the publication of Baek et al. [1]. While the most widely used fabrication technique for formation of NiO has been reactive physical vapor deposition, for scaled RRAM integration more advanced and conformal chemical vapor deposition techniques are required. We report on the unipolar switching properties of NiO synthesized by Metal-Organic Chemical Vapor Deposition technique (MOCVD) , and the effects of process parameters as NiO thickness, of Ti-doping, and of electrode thickness on the switching performance is discussed.

NiO DEPOSITION

NiO and Ti-doped NiO films were deposited by MOCVD on Ni coated 300mm Si substrates (20nm PVD Ni on top of 40nm PVD TiN/Si), using $Ni(dmamb)_2$ and Ti(TDMAT) precursors and O_2 in a MOCVD chamber on a TEL-TRIAS platform [2]. Deposition temperature was 315°C. Within wafer uniformity was better than 1.5%, and surface roughness less than 1nm for 10nm thick films. Density of the NiO films is $6.55g/cm^3$, which is close to bulk value. The films were polycrystalline with a preferred (111) orientation from XRD. From ERD analysis, films were shown to be near stoichiometric but slightly Ni rich with the Ni:O ratio of 1.1:0.9. Ti-doped NiO films up to 1.75%Ti were successfully deposited by adding Ti precursor pulses.

TEST STRUCTURE AND ELECTRICAL TEST CONDITIONS

Unipolar switching was observed on these MOCVD and ALD NiO films using Ni as top or bottom electrode and TiN as the counter electrode. For process integration reasons, we selected Ni as the uniform bottom electrode while TiN top electrode(PVD) was patterned using a conventional litho and reactive dry etching process. Top electrode size was 100um*100um.

For studying the effect of process parameters, we changed thickness of the NiO, compared undoped and Ti-doped NiO films, and investigated different thicknesses of the TiN top electrode,

DEFINITION OF GOOD SWITCHABILITY

The NiO samples show unipolar switching, with typically quite abrupt SET and RESET characteristics (Fig.1). However, the detailed switching behaviors of the different samples prepared show important differences depending on the stack thicknesses and other process parameters.

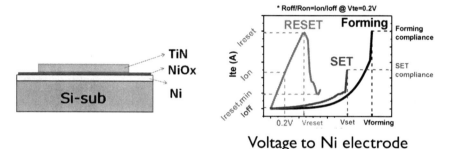

Figure 1. Schematic of test sample (left figure) and typical unipolar switching characteristics of NiO samples (schematic, right figure). Voltage needs to be positive to Ni (bottom) electrode to obtain switching.

A good switchability can be defined by limited first reset current (<1mA) and "deep" RESET ,i.e. high Roff/Ron ratio (>10) (Fig.2). It is observed that the depth of RESET (height of Roff) is depending on the current at first RESET, a higher first RESET current typically resulting in a lower Roff. This RESET current level is controlled by the maximum current during Forming [3]. Indeed, the current during Forming/Set determines the strength of the filament (Ron, or "width",[4]). In general, the filament formation can be viewed as a two step process, in which initially a conductive path is formed during a kind of dielectric breakdown, while the actual current flowing through this conductive path by Joule heating activates a thermal redox reaction (e.g.local NiO reduction to Ni) to form the actual filament.

Hence, the Forming is the most decisive step. As we are using 0T1R MIM structures, the current during the forming is not controlled by the external current compliance of the I-V generator, but by discharge of parasitic capacitances and self-capacitance [3,5]. This additional discharge current is responsible for the Reset current being higher than the externally defined compliance current (Fig. 3)

The I-V conditions at forming are further constrained by (i) the forming voltage, as a higher forming voltage results in a higher charge on the parasitic capacitances, and (ii) the leakage current at forming that sets a minimum value on the external current compliance (Fig.3)

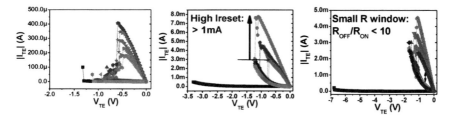

Figure 2. Typical Forming and subsequent REST/SET switching curves for (left) good quality switching cells, and poor quality switching cells characterized by a high RESET current (center) and low Roff/Ron ratio (right)

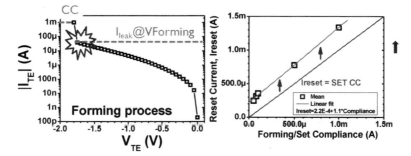

Figure 3. Limitations in the current compliance during Forming set by the leakage current at forming (left figure), and deviation of the RESET current from the compliance during Forming due to the discharge of the parasitic capacitances in the 0T1R configuration

RESULTS AND DISCUSSION

NiO thickness effect

It is observed that thinner films show better switching (Fig.4). This can be understood by (i) the lower forming voltage of thinner films (at same forming field), while (ii) at the same forming field, thinner films show a lower leakage (Fig.5) (attributed to a better dielectric quality of thinner films), so that a lower functional current compliance can be applied. This results in a less strong ("thick') filament creation requiring also a lower RESET current to break.

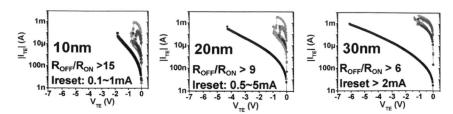

Figure 4. Forming and subsequent REST/SET switching curves for 10nm (left) , 20nm (center) and 30nm (right) NiO films. Lower RESET current and higher Roff/Ron ratio is obtained for thinner films.

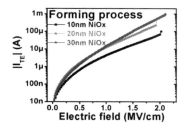

Figure 5. Thinner NiO films show a lower leakage at similar forming field.

Ti-doping effect

Ti-doped films show deeper RESET (Fig.6). Undoped and doped films show similar Forming field, however the Ti-doped film shows a lower leakage at Forming (Fig. 7), resulting in a lower functional current compliance.

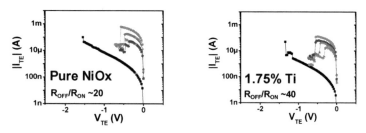

Figure 6. Forming and subsequent switching curves of undoped NiO (left) and 1.75% Ti-doped NiO films (right). Doped films shows a higher Roff/Ron ratio.

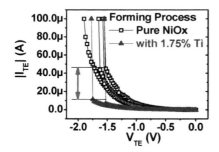

Figure 7. Leakage current of undoped and doped NiO films during forming.

PVD TiN TE thickness effect

At same NiO film thickness, a better switching is observed for thicker TiN TE (Fig.8). While forming voltage and leakage current at forming are similar, this can be explained by increased dissipation of the Joule heat during forming, so that at a same forming current a thinner filament is formed (Fig.9).

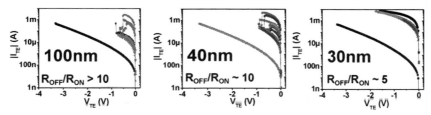

Figure 8. Forming and subsequent RESET/SET switching curves for samples with different TiN top electrode thiskness (8nm NiO film in all 3 samples)

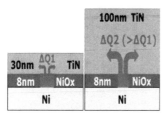

Figure 9. Schematic of higher heat drain effect of thicker Top Electrode.

CONCLUSIONS

Good unipolar switching was obtained in MOCVD NiO films using Ni bottom and TiN top electrode. The effect of structural stack variations as NiO thickness and TiO doping and TiN thickness on the switching behavior of NiO could be explained by the effect on the forming current and voltage level, and Joule heating dissipation. Thinner NiO films, Ti doping, as well as thicker top electrode improved the switching quality by decreasing the RESET current and increasing the Roff/Ron ratio. For best samples, first RESET currents as low as ~100uA and Roff/Ron ratio of 40 can be achieved even on large size (100um*100um) test structures.

REFERENCES

1. I. G. Baek, M. S. Lee, S. Seo, M. J. Lee, D. H. Seo, D.-S. Suh, J. C. Park, S. O. Park, H. S. Kim, I. K. Yoo, U-In Chung, J. T. Moon, *IEDM Tech. Dig.* 587 (2004)
2. J. Meersschaut, M. Toeller, M. Schaekers, X. Wang, B. Brijs, D. Wouters, M. Jurczak, L. Altimime, S. Van Elshocht, E. Vancoille, *Physics and Technology of High-k Materials 8, ECS Transactions* **33** (3), 313 (2010)
3. K. Kinoshita, K. Tsunoda, Y. Sato, H. Noshiro, S. Yagaki, M. Aoki, Y. Sugiyama, *Appl. Phys. Lett.* **93**, 033506 (2008)
4. F. Nardi, D. Ielmini, C. Cagli, S. Spiga, M. Fanciulli, L. Goux, D. J. Wouters, *Proc. IEEE International Memory Workshop*, 66 (2010)
5. L. Goux, J. G. Lisoni, X. P. Wang, M. Jurczak, D. J. Wouters, *IEEE Trans.Electron Devices* **56**, 2363 (2009)

Mater. Res. Soc. Symp. Proc. Vol. 1337 © 2011 Materials Research Society
DOI: 10.1557/opl.2011.983

Understanding the Role of Process Parameters on the Characteristics of Transition Metal Oxide RRAM/Memristor Devices

Branden Long, Yibo Li, and Rashmi Jha
Department of Electrical Engineering and Computer Science
University of Toledo, Toledo, Ohio 43606, U.S.A.

ABSTRACT

In this report, we studied the role of the oxygen concentration in TiO_x layer of $Ni/TiO_x/TiO_2/Ni$ stack based 2-terminal resistive random access memory (RRAM) devices. The sample with oxygen deficient TiO_x layer showed Schottky diode type J-V characteristics in the as-fabricated state while the sample with higher oxygen content in TiO_x demonstrated MIM or back-to-back connected diode behavior. The Capacitance-Voltage (C-V) profiling was performed and doping density vs. depletion width characteristic was obtained. The conductance technique was implemented to study the interface state density. The RRAM type switching behavior of these samples was studied. The sample with high oxygen in TiO_x showed filament based switching after electroforming while the sample with low oxygen in TiO_x showed switching governed by the charge trapping.

INTRODUCTION

Transition metal oxide (MOx) based 2-terminal (2-T) resistive random access memory (RRAM) /memristor devices hold tremendous potential for enabling massively dense non-volatile memory elements on chip [1,2]. In spite of recent research progress in this area, there are several unanswered questions of fundamental nature that need to be addressed. For example, the resistive switching mechanism has been predominantly explained by two broadly accepted theories: (i) formation of filaments in MOx, and (ii) movement of oxygen vacancies (V_o^{2+}) in MOx [1,2]. However, the role played by the process parameters such as the oxygen concentration in the MOx in governing the switching due to either of these phenomena is largely unknown. Furthermore, our understanding on the types of defects in these devices and their impact on governing the device performance has been very limited. In this paper, we report our work towards understanding the role played by oxygen concentration in the MOx stack in governing the RRAM/memristor switching behavior. Furthermore, we have implemented admittance spectroscopy techniques to understand the doping distribution and defects in 2-T RRAM devices based on MOx. We believe that these understandings will be critical for the development of the appropriate MOx materials, process conditions, and defect passivation techniques for the successful realization of RRAM/memristor devices.

EXPERIMENT

Sample Fabrication

We performed our studies on two sets of samples with different concentration of oxygen in the MOx stack. Figure 1 shows the schematic representation of our devices. 2-T devices consisting of $Ni/TiO_x/TiO_2/Ni$ stacks were fabricated on p-Si substrate. All materials were deposited using RF magnetron sputtering. The concentration of oxygen in TiO_x and its thickness was varied between the two samples while all the other materials in the stack remained the same. The samples were fabricated by first depositing 100 nm of Ni on p-Si and patterned using a shadow mask to define the bottom electrode (B.E.). Thereafter, TiO_x/TiO_2 stack was deposited using reactive sputtering of Ti in oxygen at a substrate temperature of 300°C. The sample MR81 was fabricated by depositing 110 nm of TiO_x with 10% of oxygen flow in argon (Ar) during reactive sputtering while sample MR83 was fabricated by depositing 90 nm of TiO_x with 8% of oxygen flow in Ar during reactive sputtering. A 25 nm of TiO_2 layer in both the samples was deposited with 20% of oxygen flow in Ar during reactive sputtering. The TiO_x/TiO_2 stack in both the samples was deposited without breaking the vacuum for a tight process control. Finally, 100 nm of Ni was deposited and patterned using a lift-off technique to define the top electrode (T.E.).

Electrical Characterization

The electrical characterization was accomplished by probing the samples in Lakeshore cryogenic probe station and measuring the current-voltage (I-V), capacitance-voltage (CV), and admittance spectroscopy (AS) characteristics using a Keithley 4200 Semiconductor Characterization System. All measurements were carried out at a vacuum pressure of 7.6×10^{-5} Torr in the probe station chamber in order to limit the interaction of ambient oxygen with the device during the measurement. In all of our measurements, the top electrode (T.E.) was biased while the bottom electrode (B.E.) was grounded as indicated in figure 1. All devices tested in this paper were 100 μm x 100 μm devices, except for the MR81 J-V switching which was done with a 200 μm x 200 μm device.

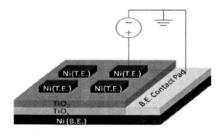

Figure 1: Schematic diagram of Ni/TiOx/TiO2/Ni RRAM device. The MR81 sample was fabricated with 110 nm of TiO_x with 10% of oxygen in Ar during reactive sputtering while MR83 sample was fabricated with 90 nm of TiO_x with 8% of oxygen in Ar during reactive sputtering. All other materials in the stack remained the same between the samples.

RESULTS AND DISCUSSION

Figures 2(a) and 2(b) shows the current density vs. voltage (J-V) characteristics in the DC voltage sweep range of -1V to 1V on as-fabricated MR83 and MR81 samples, respectively. These initial J-V characteristics on as-fabricated samples were repeatable across the entire 2" wafer indicating an excellent uniformity in the deposition conditions. Diode like behavior of the devices can be clearly observed from these figures. MR83 sample showed two orders of magnitude higher J in forward bias (at 1V) compared to the reverse bias (-1V). MR81 sample showed switching in the opposite direction compared to the MR83 sample with forward J(at -1V) to reverse J (at 1V) ratio of 50. Furthermore, orders of magnitude reduced J in MR81 sample compared to MR83 sample can be clearly observed. Since the TiO_x/TiO_2 stack in both the devices were sufficiently thick, we believe that the increased conductance of as-fabricated MR83 samples over MR81 sample can be explained by two mechanisms: (i) increased trap assisted conduction, such as Frenkel-Poole conduction or trap assisted tunneling in MR83 over MR81 due to more number of V_o^{2+} concentration in MR83 TiO_x, and (ii) lower B.E./TiO_x contact resistance in MR83 compared to MR81 due to a higher V_o^{2+} concentration in MR83 TiO_x. Based on these J-V, we modeled the MR83 samples with Ohmic contact at B.E./TiO_{2-x} interface and Schottky diode at TiO_2/T.E interface, as shown in the inset of figure 2(a). The barrier height of Ni on TiO_2 for this sample was calculated to be 0.82 eV using J-V extrapolation technique [3]. The MR81 sample, on the other hand, can be modeled using two back-to-back connected Schottky diodes due to the formation of Schottky contacts at both top and bottom interfaces. This is shown in the inset of figure 2(b). The oxidation of the bottom Ni electrode during TiO_x deposition is possible to a higher extent for MR81 samples because the TiO_x layer in this case was deposited in higher oxygen flow. If this is the case then NiO_x/TiO_x interface will lead to the formation of p-n junction, which will further attest our back-to-back connected diode model.

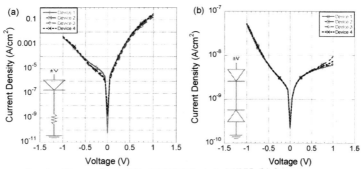

Figure 2: Current density (J) vs. voltage (V) plots: (a) Sample MR83 (b) Sample MR81.

The V_o^{2+} are known to act as donor dopants in TiO_x [4]. Therefore, to quantify and study the distribution and concentration of V_o^{2+} doping in TiO_x/TiO_2 stacks, high frequency C-V measurements were performed on the devices. Figure 3(a) shows C-V and $1/C^2$ vs. V characteristics of MR83 sample at 1 MHz ac frequency in reverse bias. The diode like C-V profile can be observed for MR83 sample. From the C-V profile, the doping density and depletion width was calculated [5]. The doping density (N_D) vs. depletion width (W) relation, assuming a dielectric constant of 30 for the TiO_x/TiO_2 stack, is shown in figure 3(b). However, the dielectric constant of TiO_2 has been reported to vary with the process conditions which has been ignored in our calculations [6]. The W is calculated from the T.E./TiO_2 Schottky interface. The doping profile was obtained between the W of 100 nm to 140 nm. The W smaller than 100 nm led to the device turn-on and therefore could not be analyzed. However, between 100 nm to 140 nm away from the T.E./TiO_2 interface, an increase in V_o^{2+} concentration from 1×10^{17} cm^{-3} to 5×10^{18} cm^{-3} was observed. Figure 3(c) shows the C-V profile of MR81 sample at 1 MHz ac frequency. A flat C-V was observed for this sample indicating Metal-Insulator-Metal type characteristic with TiO_x/TiO_2 stack in this case being devoid of V_o^{2+} and behaving like a dielectric. This observation agreed well with J-V characterization of this sample indicating significantly reduced conductance in figure 2(b). These observations clearly attested that the V_o^{2+} concentration in these two samples were significantly different as was targeted from the TiO_x process engineering.

The interface state density (D_{it}) in these samples were studied using conductance technique. In this measurement, the ac conductance (G_m) of the samples was measured at different frequencies (ω) in the range of 1 KHz to 10 MHz at different reverse bias voltages. The G_m/ω vs. ω plots for samples MR83 and MR81 are shown figures 4(a) and 4(b) respectively. D_{it}

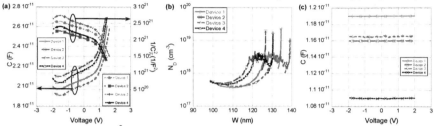

Figure 3: (a) C-V and $1/C^2$ vs. V plot for MR83, (b) Doping density (N_D) vs depletion width (W) plot for MR83, (c) C-V plot for MR81.

and time constant of defects (τ_{it}) were calculated from the maxima of G_m/ω vs. ω plots [3,7]. Clear peaks in G_m/ω vs. ω plots can be observed at different reverse bias voltages for MR83 samples in figure 4 (a). The MR81 sample also showed a peak in G_m/ω vs. ω plot, however, it was not as clear as the peaks in the MR83 sample, possibly, due to low number of defects or charge carriers in MR81 sample. Typically, D_{it} is plotted against the energy level of defects in the band-gap of the semiconductor. This energy level is obtained by calculating the surface potential (ψ_s) of semiconductor using quasi-static C-V measurement. The ψ_s obtained from C-V can be correlated with the applied bias at which D_{it} was obtained [7]. However, these relations are not well understood for the TiO_x/TiO_2 stack at this point. Therefore, we have reported the D_{it} and τ_{it} measurement at different applied voltages. Our calculations for D_{it} and τ_{it} corresponding to each maxima in the G_m/ω vs. ω plots is shown in Table I. We believe that the origin of a high concentration of D_{it} in MR83 samples is due to a high concentration of V_o^{2+} which can act as donor traps.

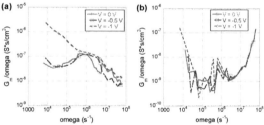

Figure 4: G_m/ω vs. ω plots for (a) Sample MR83, (b) Sample MR81. The peaks in the plot were used to calculate D_{it} and τ_{it}.

Our next sets of measurements were focused on understanding the RRAM type switching behavior of these samples. Figure 5 (a) shows the switching characteristic of MR81 sample and the inset shows the electroforming J-V sweep. The devices were formed by sweeping from 0 V to 11.7 V. The compliance current was set to 10 mA. After this forming process, device showed

Table I: D_{it} and τ_{it} calculations from figure 4(a) and (b) for sample MR83 and MR81.

	D_{it} (cm^{-2}eV^{-1})			τ_{it} (μs)		
	V = 0 V	V = -0.5 V	V = -1 V	V = 0 V	V = -0.5 V	V = -1 V
MR81	8.928E10	5.450E10	1.391E11	3.1831	3.1831	3.1831
MR83	2.0144E12	1.8678E12	1.1512E13	3.1831	1.5915	39.7887
			1.9754E12			1.0610

RRAM type switching as shown in figure 5 (a). The set voltage was observed to be -1.75V and the reset voltage was observed to be 1.45 V. The R_{OFF}/R_{ON} ratio of 228 was observed at 0.5 V read voltage. Figure 5(b) shows the switching characteristic of MR83 sample. This characteristic

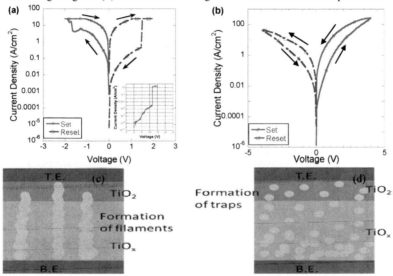

Figure 5: (a) Switching characteristic of MR81, inset shows electroforming J-V sweep, (b) Switching characteristics of MR83, (c) schematic representation of filament formation in TiOx/TiO$_2$ stack to explain filament based switching in MR81, (d) schematic representation of defects as traps in TiO$_x$/TiO$_2$ stack of MR83 to explain the observed switching behavior.

was obtained without any intentional electroforming on the sample. The switching behavior of MR83 sample was distinctly different from MR81 sample. The MR83 sample showed opposite set/reset voltage than MR81 sample as shown by arrows. The R_{OFF}/R_{ON} ratio for MR83 was observed to be 78.3 at 0.5V read voltage which was significantly lower than MR81. Based on these observations, we believe that the switching in MR81 is due to the formation of filaments.

The toggle in the set/reset current can be due to the change in the tunneling distance between T.E. and filament at set/reset voltages [8]. This model is schematically shown in figure 5(c). The formation of defect assisted filament can take place during the electroforming process due to a high electric field across the TiO_x/TiO_2 stack which behaved like a dielectric in this case. On the other hand, the switching in MR83 sample was observed due to trapping-detrapping of electrons in the V_o^{2+} traps. In order to increase the R_{OFF}/R_{ON} ratio of this sample, we attempted to electroform this sample. However, electroforming was generally unsuccessful due to a high conductivity of this sample. Therefore defect assisted filament formation failed and instead it resulted in the formation of distributed bulk traps which simply increased the overall conductivity of the sample. We have shown this model schematically in figure 5(d).

In light of these significant observations, the challenge lies in answering can RRAM type switching be achieved by introducing V_o^{2+} in TiO_x/TiO_2 stack in a controlled way through process engineering. We believe that it is possible; however, a careful understanding of the V_o^{2+} distribution in the stack, and thickness control of both TiO_x and TiO_2 layers will be critical to achieve this target.

CONCLUSIONS

In this report, we studied the role of the oxygen concentration in TiO_x layer of $Ni/TiO_x/TiO_2/Ni$ stack based 2-T RRAM devices. The sample with oxygen deficient TiO_x layer showed Schottky diode type J-V characteristics in the as-fabricated state while the sample with higher oxygen content in TiO_x demonstrated MIM or back-to-back connected diode behavior. The C-V profiling was performed and doping density vs. depletion width characteristic was obtained. The conductance technique was implemented to study the interface state density. The sample with lower concentration of oxygen in TiO_x demonstrated higher concentration of interface state density than sample with higher concentration of TiO_x. Finally, RRAM type switching behavior of these samples was studied. The sample with high oxygen in TiO_x showed filament based switching after electroforming while the sample with low concentration of oxygen in TiO_x showed switching governed by the charge trapping.

REFERENCES
1. Rainer Waser, and Masakazu Aono, *Nature Materials*, Vol. **6** (2007)
2. Joshua Yang, Matthew D. Pickett, Xuema Li, Douglas A.A. Ohlberg, Duncan R. Stewart, and Stanley Williams, *Nature Nanotechnology*, Vol. **3** (2008)
3. D.K. Schroder, 2nd Edition, Wiley-Interscience (1998)
4. Hisashi Shima, Ni Zhong, and Hiro Akinaga, *Appl. Phys. Lett.* Vol. **94**, (2009)
5. Y.Y. Proskuryakov, J.D.Major, K. Durose, V.Barrioz, S.J. Irvine, E.W. Jones, D.Lamb, Appl. Phys. Lett., Vol. **91** (2007)
6. P.Alexandrov, J. Koprinarova, T. Todorov, Vacuum, Vol.**47**, Iss. 11(1996)
7. Roman Engel-Herbert, Yoontae Hwang, and Susanne Stemmer, J. of Appl. Phys. Vol. **108** (2010)
8. Matthew D. Pickett, Dmitri B. Strukov, Julien L. Borghetti, J. Joshua Yang, Gregory S. Snider, Duncan R. Stewart, and R. Stanley Williams, J. of Appl. Phys. Vol. **106** (2009)

Mater. Res. Soc. Symp. Proc. Vol. 1337 © 2011 Materials Research Society
DOI: 10.1557/opl.2011.858

Memristive Switches with Two Switching Polarities in a Forming Free Device Structure

Rainer Bruchhaus[1], Christoph R. Hermes[1] and Rainer Waser[1,2]

[1]Peter Grünberg Institut, Forschungszentrum Jülich, 52425 Jülich, Germany and JARA –

Fundamentals for Future Information Technology, Forschungszentrum Jülich, 52425 Jülich,

Germany

[2]Institut für Werkstoffe der Elektrotechnik II, RWTH Aachen, 52074 Aachen, Germany

ABSTRACT

In this study an electroforming free device structure based on 25nm thin TiO_2 thin films is presented. The TiO_2 films are deposited on CMOS compatible W plugs. The use of 5nm thick interlayers of Ti and W between the TiO_2 and the Pt electrode turn out to be the key step to achieve the forming free performance. In these $Pt/Ti/TiO_2/W$ or $Pt/W/TiO_2/W$ samples the switching polarity can be repeatedly changed from "eightwise" to "counter-eightwise" in one device by a proper adjustment of the I-V measurement conditions. The most simple explanation for this observation is that the switching interface can be flipped back and forth from the bottom to the top electrode.

INTRODUCTION

Under the term "resistive switching" numerous approaches to use voltage induced changes of the resistance of simple two terminal metal-isolator-metal (MIM) structures for future high density data storage are subsumed. Recently, the field was reviewed and for the material systems based on nanoionic transport and redox reactions at least three different working principles were identified [1]. MIM structures based on TiO_2 with noble metal electrodes like Pt belong to the valence change mechanism (VCM). Under VCM the elemental steps of the redox reaction are anion transport and subsequent valence changes in the cation sublattice to form conducting mixed valence states or conducting phases. For TiO_2 with Pt electrodes usually the MIM structure needs an electroforming step which transforms the device from an initial high resistive state into a switchable state [2]. The subsequent bipolar switching is dominated by the metal/oxide interface and involves local changes to the electronic barrier due to the drift of oxygen vacancies in the applied electric field [3]. Different switching polarities have been observed depending on which metal/oxide interface played the key role during switching [3]. Thus interface engineering by applying different electrode materials is a fruitful approach to deepen the understanding of the switching process and the role of the interfaces. In this paper a systematic study on resistive switching of devices with W bottom electrode and three different top electrode materials is presented.

EXPERIMENT

The CMOS compatible metal W prepared as plugs embedded into a Si_3N_4-layer was used as bottom contact electrode material. The plugs had different diameters ranging from 60nm to 700nm. Prior to the TiO_2 deposition by reactive sputtering (300W, 46sccm Ar, 17sccm O_2) the W-plugs were cleaned by a short etching step for 30s in an Ar plasma. Film thickness of the TiO_2 in this study is 25nm. A set of samples with three different top electrode materials was prepared. Top electrode materials include sole 30nm Pt and double layers of 30nm Pt/5nm Ti and, finally, 30nm Pt/5nm W. Two patterning steps were performed using photoresist masks. In the first step the electrode and the TiO_2 was etched using reactive ion beam etching (RIBE) and in the second step the W contact pads were opened by etching the Si_3N_4 layer using CF_4 gas. The electrical characterization was performed under ambient conditions using an Agilent B 1500A semiconductor analyzer. Electrical signals were always applied to the top electrode while the bottom electrode was grounded.

DISCUSSION

Fig. 1a gives a schematic cross-section of the prepared device structure. The top electrode material selection includes Pt which is a noble metal with high workfunction ($\Phi_{Pt}\approx5.65eV$) and Ti and W which have considerably lower workfunction ($\Phi_{Ti}\approx4.3eV$ and $\Phi_W\approx4.6eV$) and can be oxidized much easier than Pt [4]. The Pt on top of W and Ti serves as a capping layer to suppress surface oxidation to achieve better contact with the probe tips.

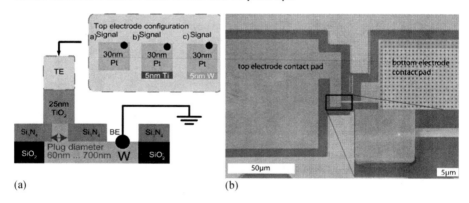

(a)   (b)

Figure 1. (a) Schematic cross sectional sample view. Top electrode (a) Pt, (b) Pt/Ti, (c) Pt/W. Electrical signals were applied to the top electrode, the bottom electrode was grounded.
(b) SEM bird's eye view on the sample structure. The inset gives a more detailed view on the plug region of the sample.

Fig. 1b is a SEM micrograph of the sample structure and, more specifically, of the W-plug region of the fabricated sample. The bottom electrode is embedded in a SiO_2 layer and a W metallization line forms the contact to the contact pad. The inset gives a more closer look to the W plug which is visible as a small light dot. The dramatic impact of the top electrode materials on the electrical properties of the devices is already readily observed by looking at the initial I-V curves, Fig. 2.

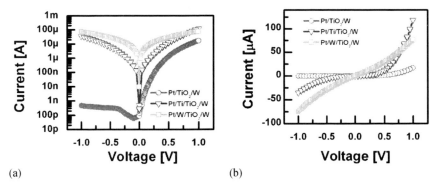

(a)   (b)

Figure 2. Initial I-V measurement of the sample with different top electrode, (a) log-linear scale, (b) linear-linear scale.

The sample with the pure Pt top electrode exhibits a rectifying behavior. Under reverse bias the current is several orders smaller compared to the samples with the Pt/Ti and Pt/W electrode. In the graph Fig. 2b where the data is plotted on a linear scale the Pt/W sample has an about linear trace whereas the sample with Pt/Ti is slightly non-linear with similar current levels. This result also indicates that depending on the selection of the top electrode material the top interfacial region dominates the electrical properties. The formation of a Schottky like barrier is thought to be the reason for the rectifying behavior. As mentioned above, Pt is a high workfunction metal and TiO_2 is a n-type semiconductor due to the doping nature of the oxygen vacancies [3]. The measurement of the initial I-V behavior needs to be restricted to about ±1V as for higher voltage amplitudes resistive switching starts to set in. Typical resistance values received from the I-V traces in Fig. 2 at -100mV are 18GΩ for the device with Pt electrode, 320kΩ and 20kΩ for the Ti and W top electrode, respectively.

The sample with the pure Pt top electrode needs an extra forming step before it can be switched. For the forming the procedure described in [5] was performed. Under positive bias voltage polarity a current controlled electroforming with steps of 100nA per 100ms was applied. A current in the range of 20 - 100μA was sufficient to trigger the irreversible transformation of the device from the highly isolating state of Fig. 1 into the switchable state. Menke et al. investigated the electroforming process in epitaxial Fe-doped $SrTiO_3$ thin films with Pt electrodes in detail [6,7]. The resistance change during electroforming was attributed to the formation of a local conductive bypass of the depletion layer at the Pt/Fe-doped $SrTiO_3$ interface. During that process oxygen is released to the environment which is considered to be electrochemically removed from the oxide film.

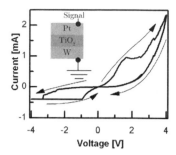

Figure 3. Typical "counter-eightwise" switching polarity of a formed Pt/TiO₂/W device.

Fig. 3 shows a typical I-V sweep of the switching of the formed device. The SET/RESET process is observed under negative/positive bias polarity, respectively, resulting in a "counter-eightwise" switching polarity. During the SET process the current needs to be controlled by a current compliance to avoid fatal damage to the device under test. Based on the observations on the switching polarity in different layer stack structures described in [3] and [5] it can be concluded that the top Pt electrode/TiO₂ interface is the switching interface. For the other two top electrode configurations Pt/Ti and Pt/W no extra electroforming process is needed to initiate the resistance switching. By comparison of the initial I-V curves and the results in [6-7] it is concluded that this advantageous feature is due to the lowered or even missing Schottky-like barrier in these samples. Most interestingly, for these devices the switching polarity can be inversed within one device under test by proper adjustment of the measurement conditions during the I-V sweep, Fig. 4.

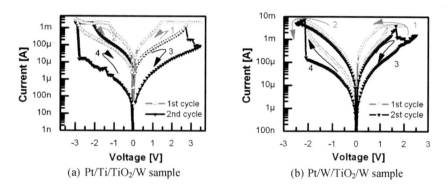

(a) Pt/Ti/TiO₂/W sample   (b) Pt/W/TiO₂/W sample

Figure 4. "Eightwise" and "counter-eightwise" switching polarity observed in the forming free device structure, (a) Pt/Ti/TiO₂/W and (b) Pt/W/TiO₂/W.

The first cycle of the I-V sweep starts from 0V with positive voltage polarity in an "intermediate" state along the loop indicated with "1". Under the control of the current compliance the device undergoes a SET process. Under negative bias voltage polarity the current compliance is removed and the sample undergoes along loop "2" a RESET process. The current compliance is kept switched off and, now along loop "3" under positive voltage a RESET process into a state with a resistance much higher than the initial "intermediate" state occurs. Finally, under negative bias voltage and along loop "4" the device undergoes a SET process which, as described above, needs to be controlled by a current compliance. The combined loops "1" and "2" are the "eightwise" switching polarity and the loops "3" and "4" are the "counter-eightwise" switching polarity. This sequence of switching polarity reversal can be repeatedly performed by a proper setting of the current compliance. For the sample with the Pt/W top electrode a very similar behavior as for the sample with Pt/Ti top electrode is observed, Fig. 4b.

Kim et al. have reported switching polarity changes observed in the symmetric Pt/TiO$_2$/Pt device structure [8]. A very complex electroforming procedure with positive bias voltage to the top electrode and application of a variable compliance current was used to vary the state of the conducting filament. After this procedure "eightwise" as well as "counter-eightwise" switching was received. In an earlier report by Jeong et al. [9] multiple electroforming steps were successfully applied to a Pt/TiO$_2$/Pt device structure that both interfaces could probably take part in the bipolar switching. Accumulation and depletion of oxygen vacancies at the active interface and subsequent redox reactions in the cationic sublattice of the oxide at the active interface is the model under discussion to explain the bipolar switching [1]. Thus, most simple explanation for different switching polarities in one device is that depending on the measurement conditions alternately the top and bottom electrode interface plays the role of the "active" interface. However, the reality might be even more complex. Muenstermann et al. report "eightwise" and "counter-eightwise" switching in epitaxial Fe-doped SrTiO$_3$ thin films on Nb-doped SrTiO$_3$ substrates after electroforming [10]. By careful analysis with conductive tip AFM it was found that "counter-eightwise" switching is related to a filamentary type of switching. The "eightwise" type was attributed to an area related switching in the halo region around the conducting filament. The filamentary "counter-eightwise" switching is easily explained by the attraction and repulsion of oxygen vacancies under the conductive tip. For the "eightwise" direction a local inversion of the doping character is assumed [10].

In summary, bipolar resistive switching turns out to be a complex interplay of defect rich oxides, different electrode materials and their barriers and the formed interfaces. These interfaces can be modified by additional forming processes which even increase the degree of freedom and complexity of the systems under investigation. Filamentary as well as homogeneous area related switching is observed. Much more systematic work is needed to fully elucidate the location of the switching and to unravel all aspects of the underlying switching mechanisms.

CONCLUSIONS

We report an electroforming-free memristive switch based on TiO$_2$ with a CMOS compatible W-plug bottom electrode and either a Pt/Ti or Pt/W tope electrode, i.e. a Pt/Ti/TiO$_2$/W or Pt/W/TiO$_2$/W layer sequence, respectively. By comparison with the Pt/TiO$_2$/W memory element it is concluded that the use of the thin Ti or W interlayer is the key step to achieve the forming free performance. These devices do not include direct high barrier Pt/TiO$_2$ interfaces and the need to bypass this barrier in an extra electroforming step has been removed. In these samples the switching polarity can be repeatedly changed from "eightwise" to "counter-eightwise" in one device by the proper adjustment of the current compliance during the I-V measurement. The most simple explanation for this observation is that the switching interface can be repeatedly flipped from the bottom to the top interface due to the reduced interfacial barriers. More detailed investigations are needed to gain a fully understanding of the observed behavior.

ACKNOWLEDGMENTS

The authors would like to thank M. Grates and R. Borowski for sample preparation and processing and T. Menke, R. Dittmann, F. Lentz and R. Rosezin for fruitful discussions.

REFERENCES

1. R. Waser, R. Dittmann, G. Staikov, and K. Szot, *Adv. Mater.* **21**, 2632 (2009).
2. J. J. Yang, F. Miao, M. D. Pickett, D. A. A. Ohlberg, D. R. Stewart, C. N. Lau, and R. S. Williams, *Nanotechnology* **20**, 215201 (2009).
3. J. J. Yang, M. D. Pickett, X. Li, D. A. A. Ohlberg, D. R. Stewart, and R. S. Williams, *Nat. Nanotechnol.* **3**, 429 (2008).
4. J. J. Yang, J. P. Strachan, F. Miao, M.-X. Zhang, M. D. Pickett, W. Yi, D. A. A. Ohlberg, G. Medeiros-Ribeiro, and R. S. Williams, *Appl. Phys. A*, published online, 26 January 2011.
5. C. Nauenheim, C. Kügeler, A. Rüdiger, and R. Waser, *Appl. Phys. Lett.* **96**, 122902 (2010).
6. T. Menke, P. Meuffels, R. Dittmann, K. Szot, and R. Waser, *J. Appl. Phys.* **105**, 066104 (2009).
7. T. Menke, R. Dittmann, P. Meuffels, K. Szot, and R. Waser, *J. Appl. Phys.* **106**, 114507 (2009).
8. K. M. Kim, G. H. Kim, S. J. Song, J. Y. Seok, M. H. lee, J. H. Yoon, and C. S. Hwang, *Nanotechnology* **21**, 305 203 (2010).
9. D. S. Jeong, H. Schroeder, and R. Waser, *Nanotechnology* **20**, 375201 (2009).
10. R. Muenstermann, T. Menke, R. Dittmann, and R. Waser, *Adv. Mater.* **22**, 4819 (2010).

Mater. Res. Soc. Symp. Proc. Vol. 1337 © 2011 Materials Research Society
DOI: 10.1557/opl.2011.985

WO$_x$ resistive memory elements for scaled Flash memories

S. Gorji Ghalamestani[1], L. Goux[1], D.E. Díaz-Droguett[2], D. Wouters[1,3], J. G. Lisoni[1,4]

[1] IMEC, Leuven, Belgium
[2] Materials Science Labs, Faculty of Physics, Pontificia Universidad Católica de Chile, Chile
[3] Electrical Engineering Dept. (ESAT), Katholieke Universiteit Leuven
[4] Physics Dept., FCFM, Universidad de Chile, Santiago, Chile

ABSTRACT

We investigated the resistive switching behavior of WO$_x$ films. WO$_x$ was obtained from the thermal oxidation of W thin layers. The parameters under investigation were the influence of the temperature (450-500 °C) and time (30-220 s) used to obtain the WO$_x$ on the resistive switching characteristics of Si\W\WO$_x$\Metal_electrode ReRAM cells. The metal top electrodes (TE) tested were Pt, Ni, Cu and Au. The elemental composition and microstructure of the samples were characterized by means of elastic recoil detection analysis (ERD), X-ray photoelectron spectroscopy (XPS), scanning electron microscopy (SEM), X-ray diffraction (XRD) and X-ray reflectivity (XRR).

Electrical measurement of the WO$_x$-based memory elements revealed bipolar and unipolar switching and this depended upon the oxidation conditions and TE selected. Indeed, switching events were observed in WO$_x$ samples obtained either at 450 °C or 500 °C in time windows of 180-200 s and 30-60 s, respectively. Pt and Au TE promoted bipolar switching while unipolar behavior was observed with Ni TE only; no switching events were observed with Cu TE. Good switching characteristics seems not related to the overall thickness, crystallinity and composition of the oxide, but on the W^{6+}/W^{5+} ratio present on the WO$_x$ surface, surface in contact with the TE material. Interestingly, W^{6+}/W^{5+} ratio can be tuned through the oxidation conditions, showing a path for optimizing the properties of the WO$_x$-based ReRAM cells.

INTRODUCTION

There has been a tremendous development in the flash memory technology over the past three decades which has introduced higher density and lower cost devices to the market. However, conventional charge based flash memories are approaching the scaling limitation of 22 nm lateral features [1]. This limitation indicates the importance of an alternative non volatile memory to replace current flash for future technology.

One alternative candidate is resistive random access memory (ReRAM), which stores the information based on resistance changes. In recent years, ReRAM has attracted much interest since it uses materials, which can be easily integrated in standard CMOS technology. Candidates for ReRAM devices are oxides where the active element can be either a binary metal oxide like NiO$_x$, HfO$_x$ and TiO$_x$ or perovskites-like compounds such as SrTiO$_3$ [2]. In this paper, we have prepared WO$_x$-based ReRAM memory cells, where WO$_x$ was obtained by the ex-situ oxidation of 50nm sputtered W films. Resistive switching is investigated using Si\W\WO$_x$\Metal structures. In particular, the role of the top electrode (TE) on the resistive switching characteristics of the WO$_x$ films was evaluated. We will show that the switching behavior correlates well with the oxide surface composition (W^{6+}/W^{5+} ratio) and not with the O:W gradient concentration found in the oxide film as it has been argued by other authors for WO_x-based ReRAM devices [3-4], where WO_x has been also obtained by the oxidation of W films.

EXPERIMENT

300 mm Si(100) wafers were used as substrate. Prior to any processing, the surface of the wafer was cleaned using HF-based aqueous solutions that remove the native oxide of the Si substrate. Then, ReRAM cell was formed as follows: W deposition, partial oxidation of the W film and TE deposition. Therefore, the Metal-Insulator-Metal (MIM) structure of our WO_x-based ReRAM cell was W-WO_x-TE. W films 50 nm thick were deposited by DC sputtering at room temperature. W oxidation process was done ex-situ via a rapid thermal annealing (RTA) in full O_2 atmosphere. The samples were oxidized at 450 °C and 500 °C during 30-220 s. TE material dots were deposited through a shadow mask. The dots diameters of the shadow mask were 30-300 μm for Pt TE and 150-1000 μm for Au, Cu and Ni. DC I-V sweep measurements used an HP4156 with 2 needles connected to different Source Monitor Units for programming and reading, respectively. Access to unoxidized W BE was done by scratching through the W\WO_x\TE film. Elemental and microstructural characterizations of memory elements were done by Elastic Recoil Detection (ERD), X-ray Photoelectron Spectroscopy (XPS), Scanning Electron Microscopy (SEM), X-ray Diffraction (XRD) and X-ray Reflectivity (XRR).

DISCUSSION

The material of the metallic TE has a strong influence on our MIM ReRAM devices, similar to has been demonstrated for other oxides such as NiO [5]. In this study, we kept fixed the BE, which corresponds to the unoxidized W film, and only the influence of different material for TE was investigated. We consider Pt as our base-line material.

Si\W\WOx\Pt electrical characterization

DC I-V sweep measurement of samples with Pt TE revealed that they have different electrical behavior that can be mainly divided into two categories: memories which did not show any switching events and memories which showed switching events. In case of non switching memories, they were either too conductive or too resistive. In the particular case of the too conductive samples, which were obtained at short annealing times, we observed that WO_x layer was thinner than 25nm. In case of too resistive samples, WO_x was thicker than 70 nm.

Table 1 DC I-V electrical measurement for Pt TE. "B": Bipolar switch; "U": Unipolar switch. Conditions marked in gray are those utilized for microstructural characterizations.

Time Temp	30 s	60 s	90 s	120 s	150 s	180 s	200 s	220 s
450 °C	No	No	No	No	No	B	No	No
500 °C	B	B(U)	No	No	No	No	No	No

Among the group of memories that displayed switching, some showed reproducible bipolar switching and others showed some partial set and reset but not reproducible results. In general, for 500 °C oxidation temperature, we observed that those samples oxidized for times longer than 90 s, Pt TE had adhering problems and could easily be removed from the oxide layer. In general, electrical observation showed that for both oxidation temperatures typically applying 2 V to Pt TE was enough to form the samples. We have summarized the electrical characterization of the samples with Pt TE in **Table 1**. An example of the bipolar switching obtained for the WO_x sample obtained at 450 °C and 180 s is shown in Fig. 1(a): initial resistance was 4 kΩ and after filament disruption (reset) and restoration (set) the cell switches between 125 Ω (high resistance state, HRS) and 30 Ω (low resistance state, LRS), respectively. The set and reset compliance current were 20 mA and 100 mA, respectively. Unipolar switching events were observed only in oxide samples obtained at 500 °C for 60 s, but the resistance window, R_{ON}/R_{OFF}, was small, i.e. <2.

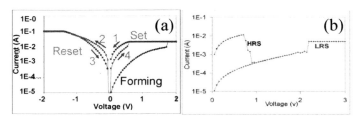

Figure 1 I-V sweep measurement of WO_x films obtained from the oxidation of W at 450°C and 180 s (a) bipolar (Pt TE) and (b) unipolar (Ni TE) switching behavior.

Effect of different TE materials

Based on the results obtained with Pt TE, we have selected some oxidation conditions to investigate the effect of TE in switching behavior: for 450 °C oxidation temperature, samples were annealed for 180 s, 200 s and 220 s and in the case of 500°C oxidation temperature, we have annealed the samples for 30s and 60 s. The materials investigated as TE were Cu, Au and Ni. A summary of the results is given in Table 2.

Table 2 Summary of switching results of different TE´s material. "B"=bipolar; "U"=unipolar.

Temperature (°C)	Time (s)	TE			
		Pt	Au	Ni	Cu
450	180	B	B	U	No
	200	No	No	U	No
	220	No	No	No	No
500	30	B	No	No	No
	60	B(U)	No	No	No

Electrical measurement with Cu TE samples did not show any switching event. Indeed, the samples were too conductive and they reached the compliance current very fast.

In case of Au TE, samples annealed at 450°C for 180 s showed reproducible bipolar switching with R_{ON}/R_{OFF}~3.5. Set compliance current and reset were 20 mA and 50 mA, respectively; the initial resistance was 400 Ω and cell switched between 70 Ω (LRS) and 250 Ω (HRS); after five cycles switching window got narrow and finally the cell got stuck in the HRS.

WO$_x$ samples obtained at 450 °C for 180-200 s and evaluated with Ni as TE showed stable unipolar switching (Figure 1(b)). Initially the cells were in ON state. By applying a reset compliance current of 50 mA, they switched to OFF state. ON state was obtained around 1-1.5 V for 180 s annealed samples and 2 V for 200 s annealed sample; setting 5 mA compliance current is enough to form the filaments, this values is 75% smaller than we normally had for the other TE's.

In summary, we have shown that different TE's affect the switching characteristics of WO$_x$ films. In particular, we have found that the initial resistance of the WO$_x$ film and forming voltage (threshold voltage) of memory cells are clearly linked to the enthalpy and work function of the metal TE of choice, respectively (Figure 2).

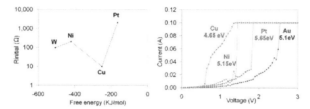

Figure 2 (a) Initial resistance as a function of the oxide enthalpy of formation and (b) DC sweep measurements of W\WO$_x$\TE cell elements. In (b) the work function values are given [6].

TE metals that are easier to oxidize than WO$_x$ induce a lower WO$_x$ initial resistance, with the only exception to Cu (Figure 2(a)). The low initial resistance when Cu is used may be a consequence of some reaction between Cu and W, reducing effectively the resistance of WO$_x$, similar to what has been observed by Kozicki *et al* [7]. Finally, as the work function of the TE metal increases thus, higher threshold voltages (first sweep) are measured (Figure 2(b)), confirming the n-type character of our WO$_x$. Metals that are easier to oxidize as compared to WO$_x$ may change locally the oxygen vacancies population at the WO$_x$-TE interface, and subsequently changing the contact resistance at that point too. In consequence, the combined effect of work function and reduction of initial resistance may induce changes in the transport of the charged species at the TE-WO$_x$ interface changing its Schottky-barrier-type characteristics, similar to what has been observed by Sawa for PCMO (p-type) and Nb:STO (n-type) materials [8]. The occurrence of bipolar (Pt and Au TE) or unipolar (Ni TE) switching may be related to the nature of the filaments formed at the TE-WO$_x$ interface, similar to what has been published for NiO [9]. This finding needs to be further investigated.

In the following, we will discuss the microstructural characterization on WO$_x$ samples formed at 450 °C for 150-180 s and 500 °C for 30-60 s. These set of samples comprehend all kind of behaviors: no switching, bipolar and unipolar switching (Table 1, gray marked results).

Elemental and microstructural characterizations

SEM characterizations showed that WO_x is smooth and uniform in thickness (Figure 3). At 500 °C, samples annealed for 30 s showed similar WO_x thickness as in the case of samples annealed at 450 °C for 150-180 s, e.g. ~35-40 nm; WO_x films prepared at 500 C for 60 s were thicker, i.e. 60-70 nm. Unoxidized W thickness was also in the range of 40 nm for the 40 nm WO_x films; for the 60 nm WO_x, W film was ~30 nm thick.

Besides of confirming the WO_x thickness as measured by SEM, XRR corroborated that WO_x films were smooth with roughness ≤3 nm. The density of the WO_x films obtained at 450 °C and 500°C/30s were in the range of 7.5(1) g/cm^3, comparable to what is reported for bulk WO_3 [10]. For samples prepared at 500°C/60s the density was smaller, i.e. 7.2 g/cm^3.

Figure 3 WO_x grown from the oxidation at 450 °C for 180 s of W films 50 nm thick.

From the crystalline viewpoint, all WO_x films were found to be tetragonal WO_3 [10]. On the other hand, the elemental composition as obtained by ERD gave us a O:W ratio ranging from 2.6 to 2.9, with no dependency on the oxidation conditions. This composition results, a sub-stoichiometric WO_3, corroborated the fact that our WO_x films are rich in oxygen vacancies. Most important, the compositions measured were uniform through the complete thickness of the oxide.

XPS surface composition revealed clearly the presence of WO_3 at the surface of all samples, corroborated by both W4f and O1s signals. XPS depth profiles showed that below the surface some other tungsten oxides with mixed valences were present (W^0, W^{+4}, W^{+5} and W^{+6}). Based on this XPS information, WO_x film can be considered as a layer-cake structure with a richest WO_3-content on the surface of the sample, similar to what has been reported for WO_x obtained by plasma techniques or thermal oxidation and used in ReRAM devices [3-4]. However, this structure may be misinterpreted since ion bombardment for XPS depth profiles analysis may reduce the oxide. Indeed, this layer-cake structure may seem contradictory with the uniform WO_x composition as measured by ERD.

Table 3 W^{6+}/W^{5+} surface ratio as measured by high resolution XPS analysis of the signal W4f.

Temperature (°C)	Time (s)	W^{6+}/W^{5+}	Switching
450	150	15.6	No
	180	11.9	B
500	30	12.2	B
	60	7.3	B(U)

High resolution XPS analysis of the W4f signal proved that the most important components of the surface of the samples are W^{6+} and W^{5+} and their ratio depended upon the oxidation conditions (Table 3), similar to what has been reported by other authors [11-12]; no W^{4+} was detected. The presence of mixed +5 and +6 valence states corroborated ERD composition evaluation. In particular, samples with no switching displayed a large W^{6+}/W^{5+} ratio ($W^{6+}/W^{5+}>15$), while bipolar switching showed an intermediate value ($W^{6+}/W^{5+}\sim12$); unipolar switching samples displayed small W^{6+}/W^{5+} ratio ($W^{6+}/W^{5+}<8$). Correlating these results to the appearance of switching events, we can conclude that WO_x compositions not close to stoichiometric WO_3 will promote the resistance switching behavior. This points out that there is a minimum amount of oxygen vacancies required for resistive switching to be triggered. Unipolar switching seems to correlate well with more defective oxide films, as represented by low density WO_x values, and larger amount of W^{+5} compounds, i.e. W_2O_5; this last observation may confirm the effect of Ni TE of reducing the WO_x film: Ni TE promoted only unipolar switching on samples where noble TE behaved bipolar.

CONCLUSIONS

We have investigated the resistive switching behavior of WO_x films obtained by the ex-situ oxidation of W films. The test vehicles for electrical characterizations were Si\W\WO$_x$\TE structures. Comparing samples with and without switching events we observed that a possible WO_x material parameter that influences the resistive switching characteritics is the WO_x surface compostion, the one in contact with the TE: sub-stoichiometric tungsten trioxide, i.e. $WO_{3-\delta}$, promotes the switching through the presence of oxygen vacancies. Unipolar switching correlates well with a high content of W_2O_5 too, but also with a more defective oxide layer. Interestingly, the surface composition of the WO_x films can be tuned through the oxidation conditions and the selection of the TE (noble metals vs. non-noble metals), showing a path for optimizing the properties of the WO_x-based ReRAM cells.

ACKNOWLEDGMENTS

We acknowledge N. Jossart for process assistance and the partial funding by IMEC's Industrial Affiliation Program on RRAM memory. JL would like to thank the financial support of Fondecyt under contract 1090332 and the Academic and Research Bureau (FCFM) for funding provided to attend the conference.

REFERENCES

1. G.I.Meijer, Science **319** (2008) 1625
2. G.W.Burr et al, IBM J.Rev & Dev. **52** (2008) 449
3. Ch.-H. Ho et al, 2007 VLSI Technology Digest of Technical Papers, 978-4-900784-03-1
4. W.-H. Ho et al, 2010 IEDM paper 19.1.1; W.C. Chien et al, 2010 IEDM paper 19.2.1
5. C.B. Lee et al, Appl. Phys. Lett **93** (2008) 042115
6. H. B. Michaelson, J. Appl. Phys **48** (1977) 4729
7. M. Kozicki et al, IEEE Trans. Nanatech. **5** (2006) 535
8. A. Sawa, Mater. Today **11** (2009) 28
9. L. Goux et al, J. Appl. Phys.**107**(2) (2010) 024512
10. JCPDS database for tetragonal WO_3 Nr 00-005-0388
11. A. Romanyuk et al, Nucl. Instr. and Meth. in Phys. Res. B **232** (2005) 358
12. R. Sohal et al, Thin Solid Films **517** (2009) 4534

Mater. Res. Soc. Symp. Proc. Vol. 1337 © 2011 Materials Research Society
DOI: 10.1557/opl.2011.1070

Memory Retention Characteristics of Data Storage Area Written in Transition Metal Oxide Films by Using Atomic Force Microscope.

K. Kinoshita[1,2], T. Yoda[1], and S. Kishida[1,2]

[1]Department of Information and Electronics, Graduate School of Engineering, Tottori University, 4-101 Koyama-Minami, Tottori 680-8552, Japan.
[2]Tottori University Electronic Display Research Center, 522-2 Koyama-Kita, Tottori 680-0941, Japan.

ABSTRACT

Conductive atomic-force microscopy (C-AFM) writing is attracting attention as a technique for clarifying the switching mechanism of resistive random-access memory (ReRAM) by providing a wide area filled with filaments, which can be regarded as one filament with large radius. We observed a C-AFM writing area of NiO films using SEM, and revealed a correlation between the contrast in a secondary electron image (SEI) and the resistance written by C-AFM. In addition, the dependence of the SEI contrast on the beam accelerating voltage (V_{accel}) suggests that the resistance-change effect occurs near the surface of the NiO film. As for the effect of electron irradiation on the C-AFM writing area, it was shown that the resistance change effect was caused by exchanging oxygen with the atmosphere at the surface of the NiO film. This result suggests that the low resistance and high resistance areas are, respectively, p-type $Ni_{1+\delta}O$ ($\delta < 0$) and insulating (stoichiometric) or n-type $Ni_{1+\delta}O$ ($\delta \geq 0$).

INTRODUCTION

Reduction/oxidation of conductive filamentary paths consisting of cationic [1] or anionic [2] vacancies are thought to be the most likely origin of a resistance change effect of ReRAM, especially of binary-transition-metal-oxide ReRAM. However, the conductive filamentary paths, termed "filaments", are sandwiched between a top and bottom electrode. Moreover, it was reported that the diameter of the filaments is very small; some groups even reported the diameter to be less than 50 nm [3-5]. Physical and chemical analyses of the filament by conventional analytical methods have therefore been hindered for a long time. Against this background, overcoming the above-mentioned difficulties by using conductive atomic force microscope (C-AFM) was reported to be possible [3,5-7]. That is, by directly contacting transition metal oxide films with an AFM-tip and scanning the tip under application of dc bias voltage, a large area with an arbitrary resistance can be formed without a top electrode being deposited. The AFM-tip plays a role as a top electrode, and the filament is formed between the tip and the bottom electrode. By scanning the AFM-tip, filaments are then generated one after another, and, as a result, the scanned area is filled with the filaments. If occupation density of the filaments in the scanned area is high enough, the scanned area can be regarded as one filament with a large diameter. This feature is a great advantage in analyses on the ReRAM filament. The scanned area will be referred to as a "C-AFM writing area", hereafter.

A C-AFM writing area is visible in a secondary electron images (SEI) of a scanning electron microscope (SEM). This enabled identification of the C-AFM writing areas from non-writing areas without etching or marking to narrow down the areas on which C-AFM writing was performed [8]. In addition, SEIs contain additional information other than the surface morphology. Dopant type (p- or n-type), dopant concentration, crystallinity, etc. also affect the

SEI contrast [9-11]. It is important to determine a parameter that decides the change in the SEI contrast caused by C-AFM writing. Elucidating this parameter is closely related to elucidating the mechanism of the resistance change.

In this paper, the dependences of SEI contrast on C-AFM writing conditions and on acceleration voltage of SEM, V_{accel}, were estimated for NiO films. In addition, the effect of EB irradiation on the size of C-AFM writing area was demonstrated. These results suggest that the resistance change caused by C-AFM writing originates in a change in carrier concentration as well as a change in carrier-type (p- or n-type) near the surface of NiO films, where interchange of oxygen with the atmosphere plays a key role.

EXPERIMENT

A 60-nm NiO film was deposited on a $Pt(100nm)/Ti(20nm)/SiO_2(100nm)/Si(625\mu m)$ substrate by DC reactive magnetron sputtering at 380 °C in mixture gas of Ar and O_2 (Ar + O_2) gases to produce a $NiO/Pt/Ti/SiO_2/Si$ (NiO/Pt) structure. Here, the SiO_2/Si structure was formed by thermal oxidation of a Si wafer and a Ni metal with the purity of 99.99% was used as a sputtering target. During the deposition, the pressure of Ar + O_2 gas and the DC power were retained at 0.5 Pa (Ar : O_2 = 0.45 : 0.05 Pa) and 1.0 kW, respectively.

For the C-AFM measurements, a commercial AFM (SPI3800N/SPA400, Seiko) was used. A Rh-coated Si tip with point diameter of 100 nm (SI-DF3-R(100nm), Seiko) was grounded, and bias voltage was applied to the Pt-BE, where BE is a bottom electrode. Low resistance state (LRS) and high resistance state (HRS) were written to the NiO film by scanning the AFM-tip with the tip contacted to the NiO film. Scanning frequency, f_{scan}, was 1.0 Hz, and the pixel number was 256×256. All C-AFM measurements were performed under ambient atmosphere.

V_{accel}-dependence of the SEI contrast of C-AFM writing areas was investigated using SEM (S4800, Hitachi), after C-AFM writing. To verify that the resistance change effect is caused due to the change in the composition of the NiO film surface between $Ni_{1+\delta}O$ ($\delta < 0$) and $Ni_{1+\delta}O$ ($\delta \geq 0$), change in resistance caused by electron beam (EB) irradiation, which generally reduces oxides [12,13], of the C-AFM writing area was also investigated.

RESULTS
Secondary electron image of C-AFM writing area

To evaluate the repeatability and reproducibility of the C-AFM writing, a fourfold overwriting process was performed as follows: Firstly, a $16 \times 16 \mu m^2$ area in the center of the writing area was written by applying writing voltage, V_{write}, of −7.0 V ("A" in Fig. 1). A 12×12 μm^2 area inside the first area was then written by applying +7.0 V ("B" in Fig. 1). After that, an inner $8 \times 8 \mu m^2$ area was written by applying −7 V ("C" in Fig. 1). Finally, a central $4 \times 4 \mu m^2$ area was written by applying +7.0 V ("D" in Fig. 1). Figure 1(a) shows a current image measured by scanning a $20 \times 20 \mu m^2$ area containing the fourfold overwriting area with reading voltage, V_{read}, of +1.0 V after the fourfold overwriting. Here, the bright and dark contrast regions correspond to LRS and HRS, respectively. HRS was written with negative bias, whereas LRS with positive bias. These results are consistent with ref. [6]. It is thus concluded that HRS can be overwritten with LRS and vice versa. It was also shown that resistance in the IRS (R_{IRS}), HRS (R_{HRS}), and LRS (R_{LRS}) satisfy the relation of $R_{HRS} > R_{IRS} > R_{LRS}$, where IRS means initial resistance state. Figure 1(b) shows a topographic image of the fourfold overwriting area that was obtained simultaneously with the current image in Fig. 1(a). A slight step between the HRS and LRS areas, where the LRS area is 0.1 nm higher on average than the HRS area, can be observed.

Figure 1(c) shows the SEI of the fourfold overwriting area observed at V_{accel} of 0.8 kV. The SEI clearly reflects the resistance: LRS and HRS are respectively observed as bright and dark areas, meaning that the secondary electron yield (SEY) for LRS is higher than that for HRS. Moreover, the IRS area, which has intermediate resistance between the LRS and HRS areas, shows intermediate brightness between the LRS and HRS areas.

Figure 1(d) shows the SEI image of the fourfold overwriting area observed at V_{accel} of 3.0 kV. In contrast to the case of V_{accel} = 0.8 kV, the brightness of the inner writing area is darker than that of the outer writing area. Moreover, in agreement with ref. [8], at V_{accel} of 3.0 kV, not only the HRS area but also the LRS area is darker than the IRS area. In all the writing processes, f_{scan} and pixel number were fixed as 1.0 Hz and 256 × 256, respectively. The narrower the writing area becomes, therefore, the longer the tip stays in contact with the film and the higher writing density per unit area becomes. In addition, the number of overwrites on the inner writing areas is higher than that on the outer areas. It was suggested, therefore, that the SEI contrast for V_{accel} = 3.0 kV reflects the number of overwrites, writing speed, and/or writing density. This also means that traces of damage to the NiO film remain after every writing process. Since the contrast between areas B and C is higher than the contrasts between areas A and B and between C and D, more severe damage is considered to be introduced when HRS is written on the LRS area than when LRS is written on the HRS area. In addition, it had been already shown that the change in SEI contrast caused by C-AFM writing is not due to an adhesion of atoms (rhodium and/or silicon) constituting the AFM tip to the writing area or desorption of carbonaceous contamination by Joule heat [8]. The V_{accel}-dependence of SEI contrast suggests that the

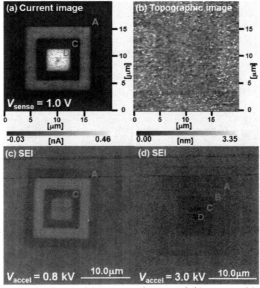

Figure 1 Simultaneously observed (a) current image and (b) topographic image of NiO/Pt film after fourfold overwriting by C-AFM. Secondary electron images (SEIs) of the fourfold overwriting area observed at V_{accel} of (c) 0.8 kV and (d) 3.0 kV.

resistance change occurred at the surface of the NiO film.

Effect of EB irradiation on C-AFM writing area

Two LRS areas were written on the same NiO film under the same conditions, namely, V_{write} = +7.0 V, f_{scan} = 1.0 Hz, and S = 20 × 20 μm^2 as shown in Fig. 2(a), where V_{write} and S are writing voltage and writing area, respectively. One LRS area is labeled area (1) and the other as area (2). Only area (1) was irradiated by electron beam (EB) with flux of 100 nA/cm² accelerated under irradiation voltage, V_{irrad}, of 3.0 kV for 5 min. Figure 2(b) shows a current image after the EB irradiation. Average current at V_{read} of 1.0 V for area (1) decreases from 71 to 25 pA, whereas average current for area (2) is almost invariant (namely, it decreases only slightly, from 96 to 94 pA). This result suggests that EB irradiation switched LRS to IRS. Figure 2(c) shows a SEI at V_{accel} = 0.8 kV containing both areas (1) and (2) after the EB irradiation to area (1) only. SEY of area (1) clearly decreases, whereas SEY of the area (2) retains bright contrast. This result confirms the correlation between SEI contrast at low V_{accel} and resistance of the C-AFM writing area.

Figures 3(a) and 3(b) show dependence of SEIs on EB irradiation time for V_{irrad} of 0.5 and 5.0 kV, respectively, where the two LRS areas were written using C-AFM under the same conditions, namely, V_{write} = +7.0 V, f_{scan} = 1.0 Hz, and S = 20 × 20 μm. Here, an index, $I_{contrast}$ = ($I_{writing}$ − $I_{initial}$)/$I_{initial}$, is introduced. This index enables numeric comparison of contrast between the writing areas and the unwritten areas, where all the pixels of the SEI are graded into 256 steps in accordance with their brightness, and the average values over the writing area and the unwritten (IRS) area are denoted as $I_{writing}$ and $I_{initial}$, respectively. The absolute value of $I_{contrast}$ becomes large when the contrast between the writing and IRS areas is strong. The contrasts disappear after 10 min and 30 min, respectively, for V_{irrad} of 0.5 and 5.0 kV under the definition that the contrast had disappeared when $I_{contrast}$ became less than 3.0. This result suggests that $I_{contrast}$ decreases faster for lower V_{irrad}. It has been reported that energy dissipation near the surface is larger for lower V_{irrad} [13,14]. It is therefore, suggested that the resistance change occurs near the surface of the NiO film, which is consistent with the V_{accel}-dependence of the SEI contrast shown in Figs. 1(c) and 1(d).

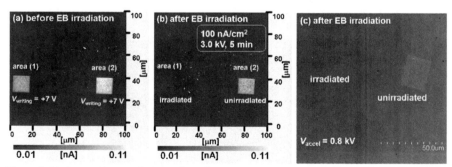

Figure 2 (a) Current image of NiO/Pt film after C-AFM writing, where two LRS areas (left and right LRS areas are denoted as areas (1) and (2), respectively) were written under the same conditions, i.e., V_{write} = +7.0 V, f_{scan} = 1.0 Hz, and S = 20 × 20 μm^2. (b) Current and (c) SE images after only area (1) was irradiated by electron beam (EB).

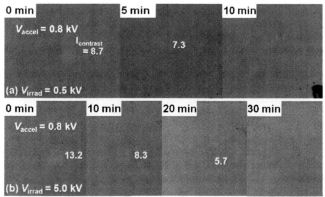

Figure 3 Electron-irradiation-time dependences of SEIs for V_{irrad} of (a) 0.5 kV and (b) 5.0 kV, where the two LRS areas were written under the same C-AFM conditions, i.e., V_{write} = +7.0 V, f = 1.0 Hz, and S = 20 × 20 μm. V_{accel} was fixed to 0.8 kV in both (a) and (b).

DISCUSSION

The contrast of the SEI for low V_{accel}, 0.8 kV, leads to the following supposition. Assuming that the LRS and HRS areas are metallic and insulating, respectively, the SEI contrast opposite to that shown by the obtained experimental image (Fig. 1(c)) should be observed because, generally, metals present lower SEY than insulators. To explain this discrepancy, it is significant to refer to the many reports about the dopant-type dependence of SEM contrast in Si [9-11]: It is well known that p- and i- or n-type areas are observed as bright and dark areas, respectively. In the same way, LRS and HRS areas written by C-AFM can be analogized as p-type and i- or n-type areas in the NiO film. It is also well known that oxides are generally reduced by EB irradiation [12,13]. It is therefore suggested that the switching of the LRS to IRS by electron-beam irradiation shown in Figs. 2 and 3 is due to reduction of the NiO surface.

Stoichiometric crystalline $Ni_{1+\delta}O$ ($\delta = 0$) is an insulator and oxygen-excess or Ni-deficient $Ni_{1+\delta}O$ ($\delta < 0$) is a p-type semiconductor. If excess oxygen is introduced as O^{2-} ion, two holes are generated as carriers in order to fulfill charge neutrality [6]. In consideration of the reduction effect of electron-beam irradiation, the switching of the LRS to IRS is considered to be brought about by reduction process of $Ni_{1+\delta}O$ ($\delta < 0$) to $Ni_{1+\delta}O$ ($\delta \geq 0$). The mechanism of the memory effect induced by C-AFM writing is therefore proposed on the basis of the results presented here as follows. When positive bias is applied to the BE, oxygen is introduced from the atmosphere into the NiO film and the surface of the $Ni_{1+\delta}O$ film becomes Ni deficient ($\delta < 0$). On the other hand, when negative bias is applied to the BE, the inverse reaction reduces $Ni_{1+\delta}O$ ($\delta < 0$) to $Ni_{1+\delta}O$ ($\delta \geq 0$). This result is consistent with ref. [7], which reported that opposite bias polarity is required to induce Ga-doped ZnO (GZO) (n-type) and NiO (p-type) into the same resistance state, simply by modifying a resistance change area in a proposed model in ref. [7] from the NiO/Pt-BE interface to the AFM-tip/NiO (ZnO) film interface. It is worth noting that the composition of the HRS area may even become n-type ($\delta > 0$), since brightness of the HRS area is darker than that of the IRS area as shown in Fig. 1(d). Lee *et al.* also reported that HRS and LRS written by C-AFM can be interpreted as oxygen vacancy formation and oxygen incorporation at the AFM-tip/NiO interface. This interpretation is based on the results of measurement of a surface-potential image that was obtained simultaneously with a current image by Kelvin-probe microscopy [15].

The reason that the resistance change at the surface determines the whole resistance in the vertical direction remains unknown. As for determining the reason, it was revealed by C-AFM and absorption current-mapping measurement (data not shown) that grain boundaries of NiO films are conductive even in HRS. It is therefore considered that a grain boundary works as a current path connecting the film surface and the BE.

CONCLUSIONS

It was shown that the combination of C-AFM writing and SEM can provide powerful method for analysing ReRAM filaments by enabling three-dimensional resistance-visualization. Specifically, it was found that resistance-change effect induced by C-AFM writing is brought about by exchanging oxygen with the atmosphere at the surface of the NiO film. That is, low-resistance state (LRS) is brought about by introducing oxygen from the atmosphere into the surface of the NiO film at the AFM-tip/NiO film interface; on the contrary, high-resistance state (HRS) is brought about by releasing oxygen to the atmosphere from the surface of the NiO film at the AFM-tip/NiO film interface. As a result, LRS area is p-type $Ni_{1+\delta}O$ ($\delta < 0$), whereas the HRS area is insulating (stoichiometric) or n-type $Ni_{1+\delta}O$ ($\delta \geq 0$). The proposed mechanism should be analyzed in a more reliable way by utilizing Monte Carlo electron trajectory simulation, which provides the size of the interaction volume and secondary radiation products as a function of specimen and beam parameters.

REFERENCES

1. M. Kawai, K. Ito, N. Ichikawa, and Y. Shimakawa, Appl. Phys. Lett. **96**, 072106 (2010).
2. P.D. Greene, E.L. Bush, and I.R. Rawlings, Proc. Symp. on Deposited Thin Film Dielectric Materials, edited by F. Vratny (The Electrochemical Society, New York, 1969), pp. 167–185.
3. K. Szot, R. Dittmann, W. Speier, and R. Waser, Phys. Status Solidi **1**, R86 (2007) (RRL).
4. M-J. Lee, S. Han, S-H. Jeon, B-H. Park, S. Kang, S-E. Ahn, K-H. Kim, C-B. Lee, C-J. Kim, I-K. Yoo, D-H. Seo, X-S. Li, J-B. Park, J-H. Lee, and Y. Park, Nano Lett. **9**, 1476 (2009).
5. H. Shima, F. Takano, H. Muramatsu, M. Yamazaki, H. Akinaga, and A. Kogure, Phys. Status Solidi **2**, 99 (2008) (RRL).
6. C. Yoshida, K. Kinoshita, T. Yamasaki, and Y. Sugiyama, Appl. Phys. Lett. **93**, 042106 (2008).
7. K. Kinoshita, T. Okutani, H. Tanaka, T. Hinoki, K. Yazawa, K. Ohmi, and S. Kishida, Appl. Phys. Lett. **96**, 143505 (2010).
8. K. Kinoshita, T. Makino, T. Yoda, K. Dobashi, and S. Kishida, J. Mater. Res. **26**, 45 (2011).
9. C. P. Sealy, M. R. Castell and P. R. Wilshaw, J. Electron Microscopy **49(2)**, 311 (2000).
10. D.D. Perovic, M.R. Castell, A. Howie, C. Lavoie, T. Tiedje, and J.S.W. Cole, Ultramicroscopy **58**, 104 (1995).
11. D. Venables, H. Jain, and D. C. Collins, J. Vac. Sci. Technol. **B 16(1)**, 362 (1998).
12. S. J. Randolph, J. D. Fowlkes, and P. D. Rack, J. Appl. Phys. **98**, 034902 (2005).
13. A. Rar and M. Yoshitake, Jpn. J. Appl. Phys. **39**, 4464 (2000).
14. H. Nonaka, S. Ichimura, K. Arai, and C. L. Gressus, Surf. Interface Anal. **16**, 435 (1990).
15. M.-H. Lee, S.-J. Song, K.-M. Kim, G.-H. Kim, J.-Y. Seok, J.-H. Yoon, and C.-S. Hwang, Appl. Phys. Lett. **97**, 062909 (2010).

Mater. Res. Soc. Symp. Proc. Vol. 1337 © 2011 Materials Research Society
DOI: 10.1557/opl.2011.986

A Survey of Metal Oxides and Top Electrodes for Resistive Memory Devices

S.M. Bishop[1], B.D. Briggs[1], K.D. Leedy[2], S. Addepalli[1], and N.C. Cady[1]
[1]College of Nanoscale Science and Engineering, University at Albany (SUNY), Albany, NY 12203, U.S.A.
[2]Air Force Research Laboratory, 2241 Avionics Circle, Dayton, OH, 45433, U.S.A.

ABSTRACT

Metal-insulator-metal (MIM) resistive switching devices are being pursued for a number of applications, including non-volatile memory and high density/low power computing. Reported resistive switching devices vary greatly in the choice of metal oxide and electrode material. Importantly, the choice of both the metal oxide and electrode material can have significant impact on device performance, their ability to switch, and the mode of switching (unipolar, bipolar, nonpolar) that results. In this study, three metal oxides (Cu_2O, HfO_x, and TiO_x) were deposited onto copper bottom electrodes (BEs). Four different top electrode (TE) materials (Ni, Au, Al, and Pt) were then fabricated on the various metal oxides to form MIM structures. Devices were then characterized electrically to determine switching performance and behavior. Our results show that the metal TE plays a large role in determining whether or not the MIM structure will switch resistively and what mode of switching (unipolar, bipolar, or non-polar) is observed.

INTRODUCTION

Metal oxide based resistive memory devices are characterized by an initial high resistance state (HRS) that can be modified to a low resistance state (LRS) by application of a characteristic threshold voltage. The mechanism of this switching behavior is being actively studied by a number of research groups and appears to vary based upon device architecture and the material(s) chosen. Such devices have been fabricated using a wide array of metal oxide insulator/metal electrode combinations. To date, few studies have rigorously compared devices fabricated from multiple electrode/insulator combinations. Those studies that have compared devices with different electrode/insulator combinations show that the choice of electrode plays a large role in the switching characteristics of the resulting MIM devices. Kim, *et al.* used several different metal TEs for TiO_x–based devices, including Pt, Au, Ni, Al, and Ti [1]. When using Pt and Au electrodes, both bipolar and unipolar switching was observed, but Ag devices exhibited only bipolar behavior. Further, devices fabricated with Ti TEs were not able to be switched, while Ni and Al TEs resulted in unstable switching behavior. Similar trends were observed by Vallee, *et al.* for HfO_x-based devices in which the switching behavior for Pt TEs was superior to Au and $WSi_{x(x>2)}$-based TEs [2]. Another study using HfO_x-based devices showed that the conductivity of the low resistance state (LRS) was related to the choice of TE material, and the heat of formation for oxidation of that material [3].

Clearly, the choice of metal oxide and electrode materials for resistive memory devices is important for their resulting electrical behavior. In this study, we chose to investigate several metal oxide/electrode combinations in an attempt to gain a broader understanding of their effects on resistive switching behavior. Cu_2O, HfO_x, and TiO_x oxides were formed on copper BEs, followed metallization with four different TEs (Ni, Au, Al, and Pt). Current-voltage measurements were then performed to characterize the switching behavior (unipolar, bipolar, non-polar), resistance ratio (R_{OFF} vs. R_{ON}) and approximate set/reset voltages.

EXPERIMENTAL PROCEDURE

The substrates for this work consisted of a 1 µm thick electroplated copper thin film on top of Ta/TaN/SiO$_x$/Si$_x$N$_y$/Si. Utilizing 300 mm wafer processing a starting substrate of SiO$_2$/Si$_x$N$_y$/Si was fabricated using standard chemical vapor deposition techniques. Atop the starting substrate Cu/Ta/TaN was deposited by physical vapor deposition (PVD) as the electroplating seed, adhesion layer, and diffusion barrier, respectively. To form the Cu BE, 1 µm of electroplated Cu was deposited onto the Cu seed layer. Chemical mechanical planarization (CMP) was then used to level and polish the plated Cu, smoothing the electrode surface. This Cu film served as the BE for all devices.

Three different metal oxide films were synthesized onto the Cu BE using thermal oxidation, physical vapor deposition and atomic layer deposition (ALD). **1) TiO$_x$ films** (100 nm thick) were deposited by RF sputtering at 200W forward power on an unheated chuck: a) in a 4.2 mTorr argon atmosphere and b) in a 4.2 mTorr argon atmosphere with an 0.1% oxygen partial pressure. These process conditions yielded a nominal deposition rate of 0.08 nm/sec. **2) HfO$_x$** films were deposited by ALD with a chuck temperature of 250 °C and a chamber pressure of 0.19 torr. Process gases used were tetrakis(dimethylamido)- hafnium(IV) as the metal-organic precursor, and a 300 W RF O$_2$ plasma as the reactant. The target thickness of HfO$_x$ was 50 nm, which required 603 ALD cycles, totaling 6.23 hrs of deposition time. **3) Cu$_2$O** films were synthesized by thermal oxidation of the copper electrode at 200-400°C in air at atmospheric pressure.

TEs were patterned using either a shadow mask or a conventional photolithography-based lift-off process. Au, Ni, Al, or Pt were deposited individually by electron beam evaporation to a final thickness of 100 nm. The resulting contacts ranged in size from 25-100 µm. For Cu$_2$O devices, "mesa" structures were fabricated by removing the excess Cu$_2$O that was not covered by the TEs. This was done through acid-based wet chemical etching of the oxide. Following fabrication, electrical characterization was performed with sweep mode current-voltage (IV) measurements. All devices were tested by biasing the TE and grounding the BE. During forming and set processes, a current compliance of 1 mA was employed. R_{OFF} vs. R_{ON} values were measured in the linear portion of the IV sweep (before hitting current compliance) and approximate set/reset voltages were determined by averaging multiple (>3) on/off cycles for each device.

RESULTS AND DISCUSSION

TiO$_x$ devices exhibited a wide range of electrical characteristics which were dependent upon both the TiO$_x$ deposition method used and the TE material. All TiO$_x$ devices required a forming voltage of <10 V to initiate the resistive switching behavior. In general, TiO$_x$ deposited in an Ar-only atmosphere yielded devices with more repeatable switching behavior than TiO$_x$ deposited in an Ar/O$_2$ atmosphere. In particular, devices deposited in an Ar/O$_2$ atmosphere with Al and Ni TEs exhibited diode-like behavior and could not be switched from HRS to LRS. Devices fabricated with Au and Pt TEs yielded the most consistent performance for both the Ar and Ar/O$_2$ deposition conditions. These devices had similar resistive switching properties with set voltages of ~0.7 V, reset voltages less than -0.4 V, and R_{OFF}/R_{ON} ratios of ~10^6. Interestingly, devices with Al and Au TEs exhibited unipolar switching behavior (turn-on and turn-off in the same voltage polarity) while devices with Ni and Pt TEs exhibited bipolar behavior (set and reset with opposite voltage polarity). Characteristic IV curves from TiO_x devices with Ni and Al TEs are shown in Figure 1.

Our devices behaved differently than those described in previous studies using TiO_x as the metal oxide. Kim et al. [1] reported both unipolar and bipolar behavior for Pt and Al TEs, while we observed only unipolar behavior for Al and only bipolar behavior for Pt. This group further showed that Ni and Al TEs yielded devices with unstable switching behavior, while we had mixed results for these electrode materials, partially dependent upon the deposition conditions of the TiO_x. Some of these differences may be due to the fact that Kim's devices used Pt BEs, while Cu BEs were used in our devices. Since the BE material was not varied in our experiments, however, we cannot determine the overall effect it has on switching behavior.

A distinguishing feature of the TiO_x-based devices was contact bubbling and delamination (see inset in Figure 1, right side). This phenomenon is well documented by multiple research groups [4,5] and has been suggested to be the result of oxygen released from the bulk TiO_x. Contact bubbling occurred during the forming process and was most pronounced when Au was the TE. Ultimately, contact bubbling will limit the integration of TiO_x-based memory devices with CMOS systems if this issue is not addressed.

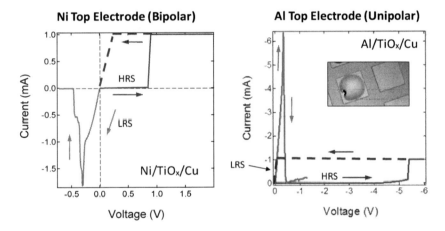

Figure 1. Current-voltage plots showing both bipolar (left) and unipolar (right) switching behavior for TiO_x devices. Bipolar behavior was observed when Ni top electrodes were used, while unipolar behavior was observed for devices with Al top electrodes. Top electrode bubbling and delamination (right, inset) was observed for all TiO_x devices, regardless of top electrode material.

HfO$_x$-based devices exhibited non-polar switching behavior independent of the TEs and without the need for a forming voltage. Non-polar behavior was exhibited by the ability to switch in both a unipolar and bipolar manner, independent of the voltage polarity for set/reset and regardless of TE material. In addition, set voltages less than +/-1 V and reset voltages on the order of +/-500 mV were observed with average R_{OFF}/R_{ON} ratios of 10^8. Examples of IV switching behavior for HfO$_x$-based devices are shown in Figure 2, below. Although this figure shows data from two different devices (one with a Ni TE and one with an Al TE) both unipolar and bipolar switching were observed for both of these TE materials.

Unipolar, bipolar and non-polar switching behavior have all been observed for HfO$_x$-based devices [2,3]. Similar to the TiO$_x$ results (above), this could potentially be due to the Cu BE used in this study. Cu may well play a role in the switching behavior that is observed for some of these devices and other work by our group has shown that Cu may diffuse to the upper surface of the HfO$_x$ during the deposition process [6]. This could have significant influence on the switching behavior of these devices.

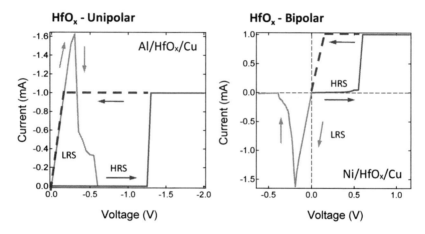

Figure 2. Current-voltage plots showing both unipolar (left) and bipoar (right) switching behavior for HfO$_x$ devices. Switching behavior for HfO$_x$ devices was independent of electrode material.

Cu$_2$O device behavior was strongly dependent upon TE material, and these devices did not demonstrate resistive switching behavior with either Au or Ni TEs. Further, only "mesa" devices exhibited consistent switching behavior. Devices that contained a continuous thermal oxide layer showed diode behavior independent of the top electrode. Mesa devices fabricated with Pt and Al TEs exhibited only bipolar switching behavior and repeated attempts at unipolar switching were unsuccessful. Current-voltage characterization indicated that the Cu$_2$O devices with Al TEs switch more consistently (at similar set and reset voltages) than those fabricated with Pt TEs. For the Cu$_2$O devices with an Al TE, set voltages ranged from 1.5-2.5 V, and reset voltages were less than -1V. No forming voltages were needed to initiate the resistive switching behavior. The R_{OFF}/R_{ON} ratio for these devices ranged from 10^3-10^4. An example IV curve is shown in Figure 3 (below) for a Cu_2O device with an Al TE.

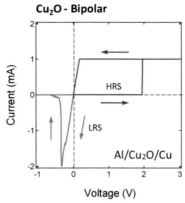

Figure 3. Current-voltage plot showing bipolar resistive switching behavior of Cu_2O based devices with Al top electrodes and Cu bottom electrodes.

In summary, devices with three types of metal oxides (Cu_2O, HfO_x, and TiO_x) demonstrated resistive switching characteristics which were highly dependent upon the type of material used for the TE. The results from this study are compiled in Table 1, below. This work shows that TE selection is an important factor for determining the resistive switching properties of MIM devices, and that this factor alone can determine switching polarity. Notably, not all TE / metal oxide combinations yielded switchable devices. Our work did not examine the role of the BE, since all devices tested used Cu BEs. Ongoing efforts in our group are focused on determining the role of BE on resistive switching characteristics, since this electrode could also have an important role in determining other aspects of switching behavior, including the set/reset voltage and R_{OFF}/R_{ON}.

Table 1. Summary of results from the metal oxide/top electrode survey.

Oxide	Synthesis	Top Electrode			
		Al	Au	Ni	Pt
Cu_xO	Thermal Oxidation	Bipolar	Diode	Diode	Bipolar
TiO_x	PVD (Ar)	Unipolar	Unipolar	Bipolar	Bipolar
	PVD (Ar and O_2)	Diode	Unipolar	Diode	Bipolar
HfO_x	ALD	Non-polar			

CONCLUSIONS

This study is one of the first to compare the effect of varying both the metal oxide insulator and the metal TE (independently) and evaluate the impact on switching polarity. Devices based on HfO_x were dramatically different from those fabricated from TiO_x and Cu_xO. Specifically, HfO_x exhibited non-polar behavior independent of the voltage polarity for set/reset and regardless of TE material. For the TiO_x and Cu_xO devices, the TE impacted the ability to resistively switch and the resulting properties if switching occurred. Another result common to both the TiO_x and Cu_xO devices was the reset process occurred when the BE was charged positively. Further work is necessary to elucidate if the charge on the bottom copper electrode plays a role on the switching behavior of our devices. The results presented in this work should help inform future studies using various combinations of materials for MIM-based resistive memory devices.

ACKNOWLEDGMENTS

This research was sponsored by the Air Force Research Laboratory awards FA87500910231 and FA87501110008. The authors would like to acknowledge the Center for Semiconductor Research at the University at Albany-SUNY for wafer development and processing and Dr. Joseph Van Nostrand, AFRL-RI for programmatic and scientific support.

REFERENCES

1. W-G. Kim and S-W. Rhee. Microelec. Eng. 87, 98-103 (2010).
2. C. Valee, P. Gonon, C. Jorel, F. El Kamel, M. Mougenot, and V. Jousseaume. Microelec. Eng. 86, 1774-1776 (2009).
3. F. El Kamel, P. Gonon, C. Vallee, V. Jousseaume, and H. Grampeix. App. Phys. Lett. 98, 023504 (2011).
4. J.J. Yang, F. Miao, M.D. Pickett, D.A.A. Ohlberg, D.R. Stewart, C.N. Lau, and R.S. Williams. Nanotech. 20, 215201 (2009).
5. D-H. Kwon, K.M. Kim, J.H. Jang , J.M. Jeon , M.H. Lee , G.H. Kim , X-S. Li , G-S. Park, B. Lee, S. Han , M. Kim and C.S. Hwang. Nat. Nanotech. 5, 148–153 (2010).
6. B.D. Briggs, S.M. Bishop, K.D. Leedy, B. Butcher, R.L. Moore, S.W. Novak, and N.C. Cady. Proc. MRS Spring 2011. (2011).

Mater. Res. Soc. Symp. Proc. Vol. 1337 © 2011 Materials Research Society
DOI: 10.1557/opl.2011.859

Switching speed in Resistive Random Access Memories (RRAMS) based on plastic semiconductor

Paulo F. Rocha[1], Henrique L. Gomes [1], Asal Kiazadeh[1], Qian Chen[1], Dago M. de Leeuw[2] and Stefan C. J. Meskers[3]
[1] Center of Electronics Optoelectronics and Telecommunications (CEOT), Universidade do Algarve, Campus de Gambelas, 8005-139 Faro, Portugal
[2] Philips Research Labs, High Tech. Campus, 5656 AE Eindhoven, The Netherlands
[3] Molecular Materials and Nanosystems, Eindhoven University of Technology, P.O. Box 513, 5600 MB Eindhoven, The Netherlands

ABSTRACT

This work addresses non-volatile memories based on metal-oxide polymer diodes. We make a thorough investigation into the static and dynamic behavior. Current-voltage characteristics with varying voltage ramp speed demonstrate that the internal capacitive double-layer structure inhibits the switching at high ramp rates (typical 1000 V/s). This behavior is explained in terms of an equivalent circuit.
It is also reported that there is not a particular threshold voltage to induce switching. Voltages below a particular threshold can still induce switching when applied for a long period of time. The time to switch is longer the lower is the applied voltage and follows an exponential behavior. This suggests that for a switching event to occur a certain amount of charge is required.

INTRODUCTION

Organic materials are becoming interesting candidates for electronic devices in new information technologies particularly on memories [1,2]. One type of memory offering excellent prospects is the resistive random access memory (RRAM), a simple diode structure whose resistance can be programmed reversibly to be high or low. In spite of intense effort, many details of the switching mechanism and charge transport have not clearly been identified. For example, there is no consensus on the write- and erase times. Reported switching times vary by orders of magnitude, the values range from nanoseconds to milliseconds. For the polymer/oxide memory diodes it was reported that short switching times (nanoseconds) are possible for individual switching events, but that repeated ON and OFF switching within a short time interval (< ms) is not possible. This phenomenon has been refered to as 'dead time' [3,4]. The origin of this effect is not known in detail. Yet it may put drastic limitation on the operation of these cells.
In this contribution we characterize the switching speed of metal-oxide memory devices using dynamic and static measurements and equivalent circuit modeling.

EXPERIMENTAL DETAILS

The organic memory used consists on a diode structure (inset of figure 1) with an Al bottom electrode, a sputtered layer of Al_2O_3 (20 nm), a spirofluorene polymer (80 nm), and a Ba/Al (5 nm/100 nm) top electrode that forms an Ohmic contact with the polymer. The devices with an active area of 9 and 1 mm^2 were encapsulated to exclude O_2 and H_2O. In all cases, positive bias voltage refers to the bottom aluminum electrode being positive with respect to the top electrode. The memories were formed; by applying a sweeping voltage from 0 to 12 V. Quasi-static current–voltage (J–V) curves were obtained using a Keithley 487 picoammeter/voltage source. The dynamic behavior of the J-V curves was recorded using a signal generator and an oscilloscope combined with a low noise amplifier. The experimental arrangement is shown in inset of figure 2. Device modeling was done using Agilent Advanced Design System (ADS).

DISCUSSION

The current-voltage characteristics are shown in figure 1. A low conductance state is designated as off-state, and a high conductance state as on-state. The most salient feature of the on-state is a decrease in current or negative differential resistance (NDR) above a certain threshold voltage, 4-5 Volts. Reliable switching is obtained at the top and bottom of the NDR [3,4]. Regarding the device represented in figure 1, using a pulse with 2–4 V amplitude, the memory switches from the high to the low resistance on-state. A pulse with 7–10 V amplitude, corresponding to the end of the NDR regime, switches the memory to the high resistance off-state.

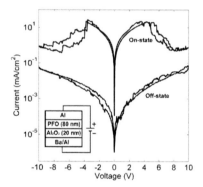

Figure 1. On-state and off-state J-V characteristics. Inset represents the device structure

The J-V characterization is measured using a triangular input wave from 0 to 10V as illustrated in inset of figure 2. In off-state the metal-oxide memory behaves as a capacitor. In the on-state, the J-V characteristics exhibit a sharp NDR as shown in figure 2. As the frequency of the input signal increases, the current decreases. The magnitude of the NDR also gradually decreases and vanishes at 1000 V/s. This effect has been previously reported by Simmons, Verderber [3] and by Verbakel et al. [4], the phenomenon has been referred to as 'dead time'.

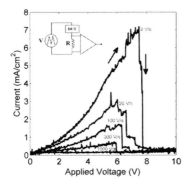

Figure 2. On-state J-V characteristics recorded at different scan rates showing the disappearance of the NDR at high scan rates. The inset shows the schematics of experimental set-up.

We propose that the reason for this counterintuitive behavior of the on-state J-V characteristics relies on the internal diode structure. The device is a double-layer comprised of a thin oxide layer (20 nm) with a high capacitance, in series with a polymer bulk region (80 nm) with a comparatively small resistance and capacitance. The response of this double-layer to a high sweeping voltage ramps is conveniently modeled by a double RC circuit. The equivalent circuit model is shown in figure 3 where C_{ox} and R_{ox} represent the oxide layer and C_{poly}, R_{poly} represent the polymer layer.

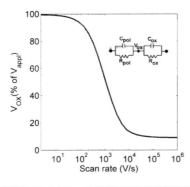

Figure 3. Internal voltage across the oxide layer as a function of the scan rate. The data was obtained using the equivalent double RC network in the inset with the following parameters: $C_{poly} = 3$ nF, $R_{poly} = 300$ kΩ, $C_{oxide} = 30$ nF, $R_{ox} = 80$ MΩ.

Previously, we have shown that this double-layer model adequately describes the frequency response of the diode [5]. Here, we will describe the behavior of this network when a ramp voltage is applied.

Figure 3 shows the internal voltage across the oxide layer as function of the voltage sweeping speed. Device parameters used were based on typical values estimated by modelling the frequency response [5]. We use $C_{poly} = 3$ nF, $R_{poly} = 300$ kΩ, $C_{oxide} = 30$ nF, $R_{ox} = 80$ MΩ. These are in line with the theoretically values estimated on geometrical parameters.

For a low ramp rate most external voltage drop is across the high impedance layer (the oxide) and only a small fraction of the external voltage is dropped across the polymer. When the sweeping frequency increases the oxide capacitance begins to be shortened and the applied voltage will rapidly increase across the polymer layer. The increase in the voltage sweeping speed has then the effect of moving the voltage drop from the oxide to the polymer layer.

The device internal structure is basically a voltage divider. However, because the oxide capacitance is much higher that the polymer capacitance, the way the applied external voltage is split between the two layers dependents on the voltage scanning speed or repetition rate. Figure 3 shows the voltage V_{ox} in percentage of V_{appl} as function of the scanning speed. Above 20 V/s, V_{ox} decreases rapidly and at 1000 V/s, 55 % of V_{appl} is now across the polymer layer. Thus, suggests that the electric field must be across the the oxide layer to observe switching.

Switching speed also depends on the voltage applied. To better understand this, the following static measurement was performed. With the device in the off-state, a voltage below the NDR region is applied until the memory switches to the on-state and the correspondent time to switch is recorded. The switching time for different voltages is shown in figure 4. The lower the voltage is, the longer the sample takes to switch to the on-state. In agreement with other authors [6], we refer to this effect as delayed-switch-on.

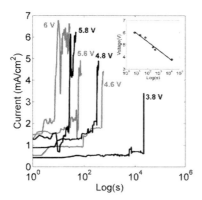

Figure 4. The device current as function of time for different constant applied biases. The inset shows that the time to switch will increase exponentially as the voltage decreases.

The sample takes 6 seconds to switch from off to on-state with an applied voltage of 6 V. However, with 3.8 V it takes 5 hours to switch. The time to switch (t_{SW}) to on-state (see inset of figure 4) follows an exponential behavior. See equation 1.

$$t_{sw} = k_0\, e^{-V_{app}/\tau}$$

(1)

Where V_{app} is the voltage applied, k_0 a constant and τ the rate of change of time to switch with the applied voltage.

CONCLUSIONS

In summary, switching speed of oxide-memory devices was discussed through static and dynamic measurements.

Non-volatile memories based on metal-oxide polymer diodes have an internal capacitive double-layer structure. The applied voltage is divided between the oxide and the polymer in a proportion that depends on the voltage scanning speed. At high scan rates the oxide impedance is shunted and the switching is inhibited. This mechanism imposes a slow dynamic response. This also confirms that switching is controlled by the oxide layer and the polymer layer is acting only as a series distributed resistance. The time to switch is also strongly dependent of the applied voltage. The mechanism behind this delayed switch remains yet to be elucidated; we may suggest a certain amount of charge must be stored within the device before switching occurs.

ACKNOWLEDGMENTS

We gratefully acknowledge the financial support received from the European project FlexNet, a Network of Excellence for the Exploitation of Flexible, Organic and Large Area Electronics, the Dutch Polymer Institute (DPI), project n.° 703, from Fundação para Ciência e Tecnologia (FCT) through the research Unit, Center of Electronics Optoelectronics and Telecommunications (CEOT).

REFERENCES

1. R. H. Friend, R. W. Gymer, A. B. Holmes, J. H. Burroughes, Marks R. N., C. Taliani, Bradley D. D. C., D. A. Dos Santos, J. L. Brédas, M. Lögdlund and W. R. Salaneck, Nature, **397**, 121-128 (1999).
2. K. S. Kwok and J. C. Ellenbogen, Moletronics, **5**, 28-37 (2002).
3. J.G. Simmons, R.R. Verderber, in *Mathematical and Physical Sciences,* (Roy. Soc. Lond. Proc. 1464, Ser. A 301 79 (1967) pp. 77-102.
4. F. Verbakel, S. C. J. Meskers, R. A. J. Janssen, H. L. Gomes, A. J. M. Biggelaar and D. M. Leeuw, Org. Electron **9**, 829-833 (2008).
5. H.L. Gomes, A.R. Benvenho, D.M. de Leeuw, M. Cölle, P. Stallinga, F. Verbakel and D.M. Taylor, Org. Electron **9**, 119 (2008).

6. M. L. Wang, J. Zhou, X. D. Gao, B. F. Ding, Z. Shi, X. Y. Sun, X. M. Ding, and X. Y. Hou, Appl. Phys. Lett. **91**, 143511 (2007).

Mater. Res. Soc. Symp. Proc. Vol. 1337 © 2011 Materials Research Society
DOI: 10.1557/opl.2011.987

Retentivity of RRAM Devices Based on Metal / YBCO Interfaces

A. Schulman[1] and C. Acha[1,2]
[1]Departamento de Física – FCEyN – Universidad de Buenos Aires, Pabellón I, Ciudad
Universitaria, C1428EHA Buenos Aires, Argentina
[2] IFIBA – CONICET

ABSTRACT

The retention time of the resistive state is a key parameter that characterizes the possible utilization of the RRAM devices as a non – volatile memory device. The understanding of the mechanism of the time relaxation process of the information state may be essential to improve their performances. In this study we examine RRAM devices based on metal / YBCO interfaces in order to comprehend the physics beneath the resistive switching phenomenon.

Our experimental results show that after producing the switching of the resistance from a low to a high state, or vice versa, the resistance evolves to its previous state in a small but noticeable percentage. We have measured long relaxation effects on the resistance state of devices composed by metal (Au, Pt) / ceramic YBCO interfaces in the temperature range 77 K – 300 K. This time relaxation can be described by a stretched exponential law that is characterized by a power exponent n = 0.5, which is temperature independent, and by a relaxation time τ that increases with increasing the temperature. These characteristics point out to a non-thermally assisted diffusion process that could be associated with oxygen (or vacancy) migration and that produces the growth of a conducting (or insulating) fractal structure.

INTRODUCTION

One of the crucial points for the development of information and communication technologies is the search of new memory devices. The Si-based devices are reaching limitations in information density, endurance and power consumption [1]. This is why a new paradigm is needed to store information in a reliably way. One of the possible candidates to overcome these technological challenges is the resistive memory devices (RRAM) based on the resistive switching (RS) mechanism, which have shown excellent properties in scalability, power consumption and operation speed, making them one of the more promising candidates to replace the actual flash memories [2-3]. One of the particular properties that must fulfill a non – volatile memory device is data retention. Here we propose to perform a detailed characterization of the lost of data retention by studying the time relaxation process after generating a switching in the remnant resistance state in $YBa_2Cu_3O_{7-\delta}$ (YBCO) / metal (Au,Pt) interfaces in order to gain insight on the physical mechanism beneath the RS on these complex oxide devices.

EXPERIMENT

In this work, we sputtered different metals (Au,Pt) on the surface of an optimally-doped ceramic YBCO sample (T_c 90 K and J_c(77 K) 10^3 A/cm^2) in order to investigate the time relaxation of the resistive states. The YBCO sample was prepared following the same procedures described elsewhere [4]. The sputtered electrodes, depicted in figure 1, have a width of 0.5 mm and a mean separation (6 ± 2) mm. They cover the entire width of one of the faces of the YBCO slab (8 x 4 x 0.5 mm^3). Finally, silver paint was used carefully to fix copper leads without contacting directly the surface of the sample.

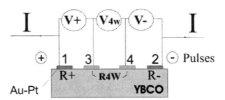

Figure 1. Contact configuration used to study the time relaxation of the remnant resistance state after producing a RS on the YBCO / metal interface. Pulses are applied on electrodes 1-2 while their remnant resistance is measured by applying a small bias current on the same electrodes and by measuring the voltage drop as indicated in the text.

Following our previous studies [5-6], at a fixed temperature (77 K < T < 320 K), we apply trains of 20,000 square pulses with an amplitude up to +5 V and 0.1 ms width at a frequency of 1 kHz to obtain a reproducible switching behavior. After that, a small bias current (~10 µA) was applied in electrodes 1 and 2 to measure different resistances using a standard DC technique; by measuring the voltage in electrodes 1 and 3 we essentially evaluate the resistance near the interface corresponding to electrode 1, as the YBCO bulk resistance between electrodes is negligible (as confirmed by measuring the four terminal resistance R_{4W}). Similarly, when we measure the voltage between electrodes 4 and 2; we essentially evaluate the resistance near the interface of electrode 2, which was always grounded. We arbitrarily call electrode 1 as R^+ (V_{13}/I_{12}) and electrode 2 as R^- (V_{42}/I_{12}).

The initial resistance of the interfaces was in the range of 20 to 50 Ω, while the bulk YBCO resistance was about 0.1 Ω for T above T_c. Thus, as already mentioned, the bulk contribution was always small or negligible. Temperature was measured with a Pt thermometer well thermally anchored to the sample, in order to detect self heating effects. When the applied power exceeded 1 W, an increase of 2 – 4 K due to self heating was observed, followed by a 10 – 20 s decrease of the temperature to its settled value. After that transient period, temperature was controlled to remain constant within a ± 0.5 K interval during all over the relaxation measurement (up to ~ 300 minutes).

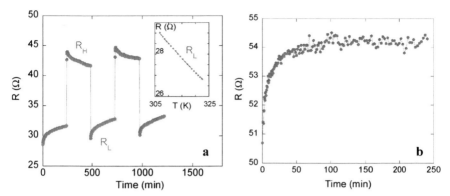

Figure 2. a) Time relaxation of the resistance state (R_H or R_L) after the switching process of the contact labeled R^+. The inset shows the typical semiconducting-like temperature dependence of the R_L state. **b)** Time relaxation of the remnant resistance state after producing a RS at T=123 K.

The time evolution of the remnant resistance of R^+ after applying the pulse train at a fixed temperature is shown in figure 2a. A similar behavior was obtained for R^-. It is clear that after each pulse train, the resistance relaxes towards its previous resistance state. Self-heating can be easily ruled out by considering the semiconducting-like temperature dependence of the resistance for each state, as shown in the inset of figure 2. If the main reason for the evolution of the resistance over time was related to this effect, then we should always observe a sudden reduction and a slow increase in resistance after applying the pulses as a consequence of the sudden increase of temperature and its slow reduction as the Joule heating dissipates, regardless of the state of resistance considered. This is not the case; we can consider that the observed evolution is related to a meta-stable state which evolves towards a more stable one. Relaxation effects were also reported in manganite-metal junctions [7] and interpreted as an evidence of oxygen diffusion in these complex oxide systems.

A detail of a single time relaxation of the remnant interface resistance after the RS with a saturation clearly defined is displayed in figure 2b. Each time relaxation can be described by the following equation:

$$\frac{R - R_0}{\Delta R} = X = 1 - e^{-\left(\frac{t-t_0}{\tau}\right)^n}$$

(1)

where R and R_0 are the resistances at time t and t_0 respectively, ΔR the total variation of R for $t \rightarrow \infty$, X the relative variation of resistance, τ the characteristic time and n an exponent. This equation is similar to the one established by Avrami [8] in order to describe the kinetics of a phase transition at a fixed temperature, where X represents the fraction of the volume that has transformed from one phase to another after a time t-t_0, and the exponent n is associated with the dimensionality of the growing process.

Equation (1) can be rewritten as follows:

$$Y = \ln[-\ln(1 - X)] = n\ln(t - t_0) - n\ln(\tau) \tag{2}$$

which can be used to fit the data and determine the n and τ parameters. As can be observed in Figure3a, the linearized data is well represented by equation (2).

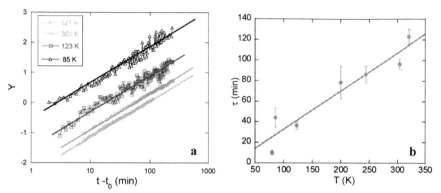

Figure 3. a) Linearization of the time evolution of the relative resistance variation by following equation (2), after producing a RS at different temperatures. **b)** Temperature dependence of the relaxation time τ in the range of 77 K to 320 K.

We have repeated the process for several temperatures in the 77 K to 320 K range; in all cases we obtained an exponent $n = 0.50 \pm 0.07$, independently of the resistance state, the temperature or the metal used for the electrode (see figure 3a). Note that the data cover almost two orders of magnitude without leaving the linear tendency. Additionally, we found that τ follows a linear relationship with T, as shown in figure 3b.

DISCUSSION

If we consider that the observed changes in resistance are linked to the evolution of a conductive phase growing over an insulating phase (or vice versa) due to a phase imbalance generated by the application of the electrical pulses, the similarity to the description of Avrami becomes natural. The obtained dimensionality (n), close to 0.5, points to a discontinuous 1D structure, like a dendritic structure with multiple cuts. A possible visualization of this structure is provided by the Feigenbaum fractal [9]. Within this framework, we can infer that the evolution to equilibrium of the phase unbalance generates a growing fractal structure in which a conductive phase is growing inside an insulator one (or vice versa) in a discontinuous form.

The fact that the relaxation time τ increases with increasing temperature is quite unexpected and clearly rules out that the growing of the fractal structure is solely assisted by thermal energy, as should be expected for a standard diffusion process. A non-monotonic temperature dependence of the relaxation time, showing a local increase of τ with increasing temperature, was also observed for other systems, like in quasi 2D monolayers [10], where the hole spin relaxation time shows a non-typical temperature behavior, that was associated with carrier interactions instead of the traditional source of scattering related to electrons and phonons, which usually dominates at high temperatures. In a similar way, our results indicate that the dominating interaction that governs the time evolution of the resistance is not a phonon-assisted process.

CONCLUSIONS

We have shown the time relaxation characteristics of the non – volatile resistance state of YBCO / metal junctions as a function of temperature. The resistance evolves to an equilibrium value towards its previous state following a stretched exponential law with a temperature independent exponent $n = 0.5$ and a relaxation time τ that increases with increasing temperature. This behavior indicates that the diffusion of oxygen (or vacancies) follows a particular law, which, by analogy with the Avrami theory, may be interpreted as an indication of the dimensionality of the diffusion path. The $n = 0.5$ exponent points out to a growing phase having a fractal structure, while the temperature dependence of τ indicates that the diffusion is not dominated by thermally assisted processes. More experimental work is needed to reveal the origin of the additional dominating process.

ACKNOWLEDGMENTS

We would like to acknowledge financial support by CONICET PIP 112-200801-00930, UBACyT X166 and CONICET-DUPONT 2010 "Memosat" Grants. We acknowledge fruitful discussions with V. Bekeris, G. Lozano, P. Levy and M. J. Sánchez and technical assistance from D. Giménez, E. Pérez Wodtke and D. Rodríguez Melgarejo.

REFERENCES

1. G. W. Burr, B. N. Kurdi, J. C. Scott, C. H. Lam, K. Gopalakrishnan, and R. S. Shenoy, *IBM J. Res. & Dev.* **52**, 449 (2008).
2. R. Waser and M. Aono, *Nature Materials* **6**, 833 (2007).
3. A. Sawa, *Materials Today* **11**, 28 (2008).
4. L. Porcar, D. Bourgault, R. Tournier, J. M. Barbut, M. Barrault, and P. Germi, *Physica C* **275**, 1997 (1997).
5. C. Acha and M. J. Rozenberg, *J. Phys.: Condens. Matter* **21**, 045702 (2009).
6. C. Acha, *Physica B* **404**, 2746 (2009).
7. Y. B. Nian, J. Strozier, N. J.Wu, X. Chen, and A. Ignatiev, *Phys. Rev. Lett.* **98**, 146403 (2007).
8. M. Avrami, *J. Chem. Phys.* **8**, 212 (1940).
9. *See, for example,* http://en.wikipedia.org/wiki/List_of_fractals_by_Hausdorff_dimension.
10. T. Li, X.H.Zhang, Y.G.Zhu, X.Huang, L.F.Han, X.J.Shang, H.Q.Ni, and Z.C.Niu, *Physica E* **42**, 1597 (2010).

Mater. Res. Soc. Symp. Proc. Vol. 1337 © 2011 Materials Research Society
DOI: 10.1557/opl.2011.981

Electro-forming of vacancy-doped metal-SrTiO$_3$-metal structures

Florian Hanzig, Juliane Seibt, Hartmut Stoecker, Barbara Abendroth, Dirk C. Meyer
Institute of Experimental Physics, TU Bergakademie Freiberg, 09596-Freiberg, Germany.

ABSTRACT

Resistance switching in metal – insulator - metal (MIM) structures with transition metal oxides as the insulator material is a promising concept for upcoming non-volatile memories. The electronic properties of transition metal oxides can be tailored in a wide range by doping and external fields. In this study SrTiO$_3$ single crystals are subjected to high temperature vacuum annealing. The vacuum annealing introduces oxygen vacancies, which act as donor centers. MIM stacks are produced by physical vapor deposition of Au and Ti contacts on the front and rear face of the SrTiO$_3$ crystal. The time dependent forming of the MIM stacks under an external voltage is investigated for crystals with varying bulk conductivities. For continued formation, the resistivity increases up to failure of the system where no current can be measured anymore and switching becomes impossible.

INTRODUCTION

Strontium titanate is a model substance for perowskite type transition metal oxides. Applications of SrTiO$_3$ range from high-k gate dielectrics to non-volatile resistance random access memories and oxygen sensors. In its initial state strontium titanate is an insulator (referring to the band gap larger than 3 eV). SrTiO$_3$ has a cubic perowskite structure, which consists of a TiO$_6$ octahedra (with the titanium on the center position of the cube) and a 12 fold O-coordinated strontium atom (on the edges). On the one hand the electronic properties can be modified by adding extrinsic dopants to the lattice. Established dopant elements are niobium, iron (on titanium lattice position), lanthanum and zirconium (on strontium lattice position) [1]. On the other hand conductivity of SrTiO$_3$ was established by high vacuum annealing introducing charged oxygen vacancies acting as intrinsic donor centers [2]. Above a threshold temperature of around 450 °C [3] the oxygen vacancy concentration depends on the surrounding oxygen partial pressure and an equilibrium concentration can be expected to be established within minutes. In this paper however, it was found that the oxygen exchange with the atmosphere takes place in a much longer time scale. MIM cells for microelectronic applications are naturally based on thin film stacks. In general, before resistance switching occurs in such MIM cells, an electro-forming at voltages larger than during normal operation are required. Obviously, the thin film real structure and the interface between insulator and metal electrodes determine the electronic transport properties. In this work, however we step back and focus on single crystal material since the correlation between the real structure and electronic properties can be separated in surface and volume contributions in a more straightforward manner than for thin film material. Therefore the current voltage (IV) characteristics of metal – SrTiO$_3$ – metal elements are investigated with emphasis on the metal – insulator – contact and changes of the current voltage (IV) characteristics introduced during electro-forming.

EXPERIMENT

Undoped strontium titanate single crystals of (100) orientation were obtained from Crystec, Berlin. The crystal dimensions are 5 x 5 x 0.1 mm^3 and 10 x 10 x 0.1 mm^3. Electrical conductivity of the SrTiO$_3$ crystals was induced by annealing in vacuum in a carbolite tube furnace at a temperature of 900 °C and a residual gas pressure of 1 x 10^{-6} mbar. The annealing time was varied from 10 hrs to 60 hrs. Metal thin films were deposited on both crystal faces by thermal evaporation to fabricate a MIM structure. To achieve a rectifying electrical behavior a noble gold contact was evaporated on the front side whereas an ignoble titanium electrode was applied on the reverse side. The measurements of the current voltage (IV) characteristics and the formation in a dc electrical field were performed with a Keithley 4200 semiconductor characterization system (SCS) in combination with a Suss Microtec test site. To confirm the results some measurements both characterization and formation were carried out using a Keithley 237 source measure unit and a single needle set up. The current compliance of both measurement devices was held constant at 1 x 10^{-1} A for the lower voltage ranges (up to 20 V) and at 1 x 10^{-2} A for the high voltage range (20 – 200 V) of the Keithley 4200-SCS. IV characteristics were measured in a range from -5 to 5 Volts. The formation process was investigated for a constant driving voltage over a time scale of hours. Table I summarizes the annealing condition in high vacuum (HV) and resulting conductivities of the samples used for forming experiments.

Table I: Properties of some exclusive strontium titanate single crystal samples.

sample index	batch	time in HV, 900 °C [hrs]	Conductivity [Ω^{-1}cm^{-1}]
1	A	10	6.6 x 10^{-09}
2	B	10	7.0 x 10^{-14}
3	B	20	3.7 x 10^{-13}
4	B	60	7.5 x 10^{-06}

DISCUSSION

Figure 1 demonstrates the effect of HV annealing at a temperature of 900 °C on the conductivity of SrTiO$_3$ single crystals as a function of annealing time. The reducing annealing increases the conductivity for nearly ten orders of magnitude with increasing annealing time. The data shown are obtained from crystals originating from the same manufacturer batch A. It should be noted that the conductivity behavior after high temperature vacuum annealing varies considerably from batch to batch. Considering the oxygen diffusion coefficient in SrTiO$_3$ at 900 °C and an equilibrium exchange of oxygen atoms from the crystal with the surrounding residual atmosphere of 10^{-6} mbar at the crystal surface, a homogeneous oxygen vacancy concentration referring to the exchange with the surrounding atmosphere [4] due to chemical diffusion inside the crystal volume is expected to be established after a few minutes [5]. A possible explanation of the increase of conductivity at a much smaller rate, that is observed here, could be a damaged region near the crystal surface that hinders the oxygen exchange with the atmosphere. Such a defect-rich near surface region is likely introduced during cutting and polishing of the crystals and is characterized by a high dislocation density and the presence of hydrogen [6].

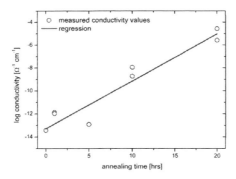

Figure 1: Logarithm of SrTiO₃ conductivity as function of annealing time in HV at 900 °C.

Figure 2a and 2b display the current measured across the titanium – SrTiO₃ – gold samples during forming in a dc electrical field for 10 hrs with the SrTiO₃ crystal previously annealed for 10 hrs and 60 hrs, respectively.

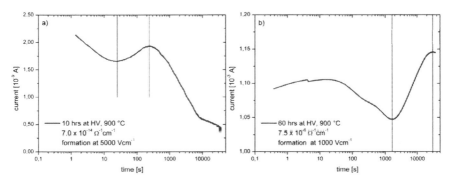

Figure 2: Current during a 10 hrs formation run with a voltage of a) 50 V and b) 10 V on a logarithmic time scale.

Due to the high conductivity a voltage of only 10 V was applied to 60 hrs annealed sample to avoid readings above the current compliance. In contrast the low-conductive, 10 hrs annealed sample was probed with a formation voltage of 50 V. Both applied voltages led to an electron as well as oxygen anion flow in the direction of the noble gold contact. Oxygen vacancies in this picture move to the ignoble titanium electrode. In Figure 2a and 2b both samples show a particular current minimum followed by a rise until a maximum value (in this order assigned by two vertical marks). In figure 3 the repeated formation in a dc electrical field of a gold – SrTiO₃ – titanium sample vacuum annealed at 900 °C for 20 hrs with a conductivity of $3.7 \times 10^{-13} \, \Omega^{-1} cm^{-1}$ is illustrated.

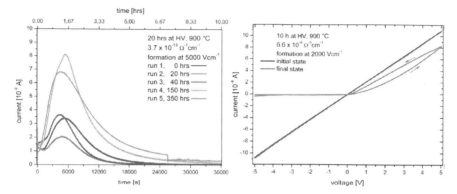

Figure 3: Current during repeated 10 hrs formation runs with a voltage of 50 V, last measured value were at all time lower than the initial records, moment of the beginning of the formation run given with respect to the start of the 1st run.

Figure 4: Electrical characteristic of a gold – $SrTiO_3$ – titanium sample changed from linear to rectifying by applying a dc electrical field (formation).

For each formation cycle the initial current values were all in the range from 1×10^{-9} to 1.5×10^{-9} A. This gives evidence of a reversible oxygen anion or oxygen vacancy movement during the formation resulting in a relaxation process. After the increasing current that is ascribed to growing flux of oxygen anions and a reverse flux of oxygen vacancies, the maximum current values slightly shift around 5000 s followed by decreasing current converging at a level lower than the initial values. Also the data presented in figure 2a show this drop of conductivity after saturation current was reached. Effectively, the increasing vacancy/anion concentration gradient counteracts further accumulation of the respective species at the electrodes. Alternatively, the peak values after a certain period of time shown in figure 3 can be interpreted as the evidence for a redistribution of oxygen ions and strontium oxide in the interface near $SrTiO_3$-region [7] of approximately 25 nm. This redistribution can create internal stress which would be an additional factor for the decreasing current values after 5000 s. In contrast, the development of conductive filaments would lead to a non reversible short circuit between both electrodes. The reversibility of the formation therefore points towards homogeneous volume conductivity without filamentary paths. The effect of the formation in a dc electrical field to the metal – $SrTiO_3$ contact is illustrated in figure 4. A distinct change from an ohmic behavior before formation to a diode characteristic afterwards is observed. The rectifying behavior is obtained after 5 forming cycles of 600 s at 50 V. Thus the formation in the dc electrical field led to a well defined Schottky contact at the $SrTiO_3$ – gold interface. This behavior could be assigned to disappearing band gap states corresponding to oxygen vacancies by the use of oxygen anion movement towards the gold - $SrTiO_3$ interface in the electrical field (schematically illustrated in figure 5).

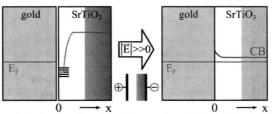

Figure 5: Band gap states in strontium titanate near the gold - SrTiO₃ interface vanish due to formation in the dc electrical field, specific Fermi energy of both materials is shown to figure out the preconditions for the contact formation, exemplarily the conduction band (CB) is sketched for the SrTiO₃ crystal after contact formation.

An alternative explanation is also based on the motion of oxygen vacancies in a dc electric field towards the cathode. Undoped SrTiO₃ single crystal samples are usually insulating or very weakly p-type conductive, since acceptors are formed by native defects such as strontium vacancies or contaminating elements like iron, which are introduced during crystal growth, or cutting and polishing. On the other hand, n-type conductivity in SrTiO₃ is composed of ionic and electronic contributions and originates from oxygen vacancies.

The introduction of oxygen vacancies by vacuum annealing at 900 °C leads with increasing vacancy concentration first to a reduction of the p-type conductivity, an intrinsic conductivity (compensation) and for oxygen-vacancy concentrations of more than 10^{19} to an n-type, electron dominated conductivity [3]. Moving oxygen vacancies in a larger dc electrical field leads to an accumulation at the SrTiO₃ – Ti contact and respectively to a depletion of oxygen vacancies at the SrTiO₃ – Au contact. This results in a linear polarization or an even more abrupt transition from very low to very high density of oxygen vacancies across the crystal between the two electrodes and hence a p to n transition from the titanium-cathode towards the gold-anode [3]. This is schematically illustrated in figure 6.

Figure 4 also shows the possibility for resistive switching with both low and high resistance states in the positive part of the IV-characteristic [8].

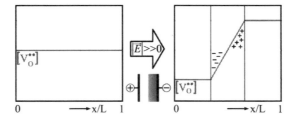

Figure 6: Constitution of an internal pn-junction due to the varying local Fermi energy in the strontium titanate crystal over the complete crystal thickness L induced by different oxygen vacancy levels.

CONCLUSIONS

In this paper the influence of the formation in a dc electrical field on single crystal strontium titanate samples in a gold – $SrTiO_3$ – titanium structure was investigated. The measured current during several formation cycle show a typical evolution with the time of the formation. We suggest that this process is closely linked to the electric field driven movement of oxygen anions and oxygen vacancies leading to a gradient of this two species in the inner crystal. Additionally, the formation leads to the establishment of a well defined Schottky contact at the $SrTiO_3$ – Au interface. This is explained with annealing of near surface trap states as well as constitution of an internal pn-junction.

ACKNOWLEDGMENTS

We thank the DFG for financial support within PAK 215. This work was performed within the Cluster of Excellence "Structure Design of Novel High-Performance Materials via Atomic Design and Defect Engineering (ADDE)" that is financially supported by the European Union (European regional development fund) and by the Ministry of Science and Art of Saxony (SMWK).

REFERENCES

1. R. de Souza, J. Fleig, R. Merkle, and J. Maier, Zeitschrift fuer Metallkunde 94, 218 (2003)
2. N. H. Chan, R. K. Sharma, and D. M. Smyth, Journal of the Electrochemical Society 128, 1762 (1981)
3. T. Baiatu, R. Waser, and K. H. Haerdtl, Journal of the American Ceramic Society 73, 1663 (1990)
4. R. Merkle, and J. Maier, Angewandte Chemie-International Edition 47, 3874 (2008)
5. H. Schmalzried, and A. D. Pelton, Solid state reactions, 2nd edition, pp. 81-85, Verlag Chemie Weinheim (1981)
6. J. Seibt, B. Abendroth, F. Hanzig, H. Stoecker, R. Strohmeyer, D. C. Meyer, S. Wintz, M. Grobosch, M. Knupfer, C. Himcinschi, U. Muehle, F. Munnik, submitted to Journal of Applied Physics (2011)
7. M. Bobeth, N. Farag, A. A. Levin, D. C. Meyer, W. Pompe, and A. E. Romanov, Journal of the Ceramic Society of Japan 114, 1029 (2006)
8. H. Stoecker, M. Zschornack, J. Seibt, F. Hanzig, S. Wintz, B. Abendroth, J. Kortus, and D. C. Meyer, Applied Physics A: Materials Science & Processing 2, 437 (2010)

Phase Change, Ferroelectric, and Organic Memories

Mater. Res. Soc. Symp. Proc. Vol. 1337 © 2011 Materials Research Society
DOI: 10.1557/opl.2011.978

Interface Characterization of Metals and Metal-nitrides to Phase Change Materials

Deepu Roy[1], Dirk J. Gravesteijn[1] and Rob A. M. Wolters[1,2]
[1]NXP Semiconductors, High Tech Campus 4, 5656AE, Eindhoven, The Netherlands
 E-mail: deepu.roy@nxp.com, phone: +31(0)402726872.
[2]MESA+ Institute for Nanotechnology, University of Twente, Enschede, The Netherlands.

ABSTRACT

We have investigated the interfacial contact properties of the CMOS compatible electrode materials W, TiW, Ta, TaN and TiN to doped-Sb_2Te phase change material (PCM). This interface is characterized both in the amorphous and in the crystalline state of the doped-Sb_2Te. The electrical nature of the interface is characterized by contact resistance measurements and is expressed in terms of specific interfacial contact resistance (ρ_C). These measurements are performed on four-terminal Kelvin Resistor test structures. Knowledge of the ρ_C is useful for selection of the electrode in the integration and optimization of the phase change memory cells.

INTRODUCTION

Embedded Phase Change Random Access Memory (ePCRAM) technology is a one transistor one resistor memory concept [1]. The resistive element is a thin PCM layer which can be switched between amorphous and crystalline states. The large resistance contrast between these two states is utilized to achieve the memory functionality. The total resistance of a ePCRAM line cell in the current path includes the resistance of the switching part of the line (R_L) and the metal-to-crystalline PCM contact resistance (R_C). For efficient cell switching at maximum power (reset pulse) the total resistance of the memory cell should match the impedance of the accompanying transistor, where R_C acts as a parasitic resistance. The larger R_C, the more the power will be dissipated at the contacts and the cell will be less efficient during switching. As the memory cell size should scale with the technology node and the relative contribution of the contacts to the cell resistance will increase. In addition, for the same technology node, the metal electrode at the contact also influences the contact resistance. Although important in switching and scaling of phase change memories, very little attempts have been performed to describe the metal-to-PCM contact properties [2].

The phase change material of choice suitable for a line cell is a fast-growth material like doped-Sb_2Te. In this work the metal to doped-Sb_2Te contact resistance is studied for different CMOS compatible electrodes materials W, TiW, Ta, TaN and TiN. This is done by contact resistance measurements on Kelvin resistor structures fabricated with these different electrodes. From these measurements ρ_C is extracted both in the amorphous and crystalline states of the PCM. The metal to PCM interface properties are described with a standard metal semiconductor model.

EXPERIMENT

The Kelvin resistor structure is a planar test structure suitable for measurement of the interfacial contact resistance with minimum parasitic resistance interference [3]. The schematic of a Kelvin resistor structure and its cross section along the contact area is shown in figure 1.

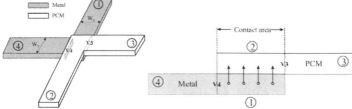

Figure 1: Schematic representation of a Kelvin resistor structure showing the metal and the PCM segments and its cross-section along the contact region.

The contact resistance measurements were performed on these Kelvin resistor structures by forcing a current from the metal-to-PCM (1 to 2) and measuring the voltage orthogonal to the direction of current flow (3 and 4). This allows measurement of the average voltage, V (equal to V3-V4) at the contact, from which the measured contact resistance, R_K is calculated. A four terminal measurement avoids probe-to-contact pad resistance and the resistance of the current and voltage taps up to the contact region. In addition these structures are designed such that least parasitic current spreading resistances are involved. The exact contact area is determined by Scanning Electron Microscope (SEM) inspection. The specific contact resistance, ρ_C is extracted from the I-V curve at a small region around zero and is expressed as:

$$\rho_C = \left(\frac{\partial J}{\partial V}\right)^{-1}_{V=0} = R_K \times A$$

(1)

To fabricate the metal to doped-Sb$_2$Te Kelvin structures, first a 100 nm metal layer is deposited by sputtering on an oxidized silicon wafer which is subsequently patterned to form the bottom electrode layer at the contact. A 500 nm PECVD SiO$_2$ layer is then deposited and the wafer surface is planarized by Chemical Mechanical Polishing (CMP) down to the metal layer. Wafers were prepared with sputter deposited W, TiW, Ta, TaN and TiN electrodes. TaN and TiN films are formed by reactive sputtering in Ar-N$_2$ ambient from there corresponding high purity metal targets. TaN deposited with a 9% partial pressure of N$_2$, results in an electrical resistivity of 2.2 $m\Omega$.cm and TiN deposited with a 40% partial pressure of N$_2$, resulted in an electrical resistivity of 240 $\mu\Omega$.cm. Process remnants and the native oxide layer on the electrode surface is removed by insitu sputter etching in argon plasma, and a 50 nm doped-Sb$_2$Te is then deposited by sputtering, resulting in amorphous layers. This PCM layer is patterned with a maximum thermal budget of 90°C to form the metal to PCM contact with an area (A). The PCM in these structures were capped with a 25 nm evaporated SiO$_2$ layer to protect against oxidation and evaporation during subsequent anneals.

The amorphous to crystalline transition temperature for this doped-Sb$_2$Te is 160 °C when heated at a ramp rate of 5°C/min. Hence in these fabricated structures the PCM is in the amorphous state and will remain amorphous when annealed at temperatures less than 150°C. The

ρ_C of metal to amorphous doped-Sb$_2$Te is extracted from the contact resistance measurements on these structures. Subsequently these structures are annealed at different temperatures in the range 175°C to 250°C. As a consequence the films are transformed into the crystalline state. The ρ_C of metal to crystalline doped-Sb$_2$Te is extracted from the contact resistance measurements on these annealed structures.

Contact resistance measurements were performed on Kelvin resistor structures with different metal electrodes. Figure 2 shows the extracted ρ_C from these measurements (using equation 1) for doped-Sb$_2$Te in the amorphous (a) and crystalline state (b).

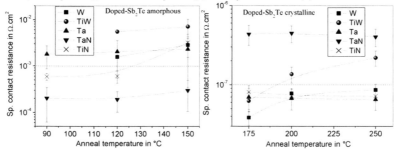

Figure 2: Change in extracted ρ_C as a function of annealing temperature in the amorphous and in the crystalline states.

The extracted ρ_C for metal to PCM depends on the state of the PCM and the metal electrode. The extracted ρ_C in the amorphous state is higher than in the crystalline state for all the metal electrodes. In the amorphous state of the doped-Sb$_2$Te, metal nitrides electrodes (TaN and TiN) have a lower ρ_C as compared to metal electrodes (Ta, TiW and W). The lowest ρ_C is extracted for TaN and the highest ρ_C for TiW. In the amorphous state the extracted ρ_C for all electrodes remain more or less constant when annealed up to 150°C. In the crystalline state, the extracted ρ_C for both metal and metal nitrides is in the same range. The highest ρ_C is measured for TaN. For TiW and W, the extracted ρ_C with PCM in the crystalline state increases when annealed at a higher temperature.

The change in extracted ρ_C with measurement temperature in the range of -40°C to 40°C for different electrode materials is shown in figure 3. These measurements are for doped-Sb$_2$Te in the amorphous state (after 120°C anneal) and in the crystalline state (after 200°C anneal).

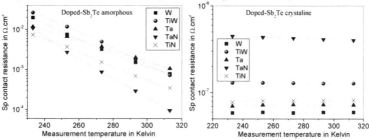

Figure 3: Change in extracted ρ_C with temperature in the amorphous and in the crystalline states.

In the amorphous state an exponential dependence is observed for ρ_C with temperature for all the electrodes, while in the crystalline state ρ_C is almost independent of the temperature.

DISCUSSION

In the amorphous state, a lower electrical interfacial resistance is observed for metal nitrides to doped-Sb₂Te as compared to the metal electrodes. In the case of pure metal electrodes, a chemical reaction is expected at the metal-PCM interface which leads to a better defined and clean interface. In general, a clean interface results in a lower interfacial resistance (ρ_C) [4]. In the case of metal nitrides, less chemical reaction is expected. The higher extracted ρ_C values for metal to PCM suggests that no barrier is created by a chemical reaction, but the electronic conduction mechanism itself is determining. The work function of metals is larger compared to the corresponding nitrides and the extracted ρ_C is also higher.

If a barrier is formed at the metal-PCM interface, the current transport through this barrier constitutes of a thermionic emission component and a tunneling component. The relative magnitude of these two limiting cases at an interface depends on the barrier height, temperature, and carrier concentration. In the case of metal-semiconductor contact, the conduction mechanism that dominates for a given carrier concentration (N_C) is determined based on the characteristic energy, E_{00}. This is a material constant that determines the tunneling probability [5] and is defined as [6] [7]:

$$E_{00} = \frac{qh}{4\pi}\sqrt{\frac{N_C}{m^*\varepsilon}} = 18.5\times10^{-12}\sqrt{\frac{N_C}{(m^*/m)\varepsilon_r}}, \qquad (2)$$

where h is Planck's constant, q is the electronic charge, m^* is the effective mass of tunneling electron, m is the free electron mass, and ε_r is the dielectric constant. A calculated value of E_{00} close to kT (which is 0.024V at room temperature) indicates thermionic field emission, a much larger value than kT indicates tunnelling at the contact (k is the Boltzmann's constant and T is the temperature in Kelvin).

In the crystalline state of doped-Sb₂Te, a carrier density of 2×10^{21} cm⁻³ and mobility of 8cm²/V.s is measured [2]. The calculated resistivity of the same PCM in the crystalline state is $36\times10^{-5}\Omega$.cm, while in the amorphous state it is 15Ω.cm. That is a four orders of magnitude change in resistivity. The carrier mobility of PCM's in amorphous state is approximately 0.1cm²/V.s [10][11]. This means that, as the PCM changes from amorphous to crystalline state the resistivity decreases by four orders in magnitude, while the mobility increases by only two orders. Hence the carrier density also increases by at least two orders in magnitude.

Using the reported dielectric constant ($\varepsilon_r\approx17.7$) for PCM in the amorphous state and 38 in the crystalline state [8] [9], E_{00} is calculated. This calculated value of E_{00} is 0.14eV in the crystalline state indicating tunneling as the major conduction mechanism, and 0.0203eV in the amorphous state indicating that thermionic field emission will be much more significant. In these calculations the tunneling electron mass, m^* is assumed equal to the electron mass, m [12]. Calculations with reported m^*/m of 0.69 in [13] results in, E_{00} of 0.16eV in the crystalline and 0.0237eV in the amorphous states and the same behavior is observed.

For PCM in the amorphous state, the extracted ρ_C changes from $\approx10^{-3}$ Ω.cm² to $\approx10^{-4}$ Ω.cm² for different electrodes (figure 2). In addition an exponential dependence of ρ_C is observed with temperature for all the electrodes. This indicates that for PCM in the amorphous state the charge transport through the metal-PCM interface is dominated by thermionic-field emission, which is dependent on the barrier height and temperature.

For PCM in the crystalline state the charge transport at the interface is dominated by tunneling and the extracted ρ_C is independent on the electrode material [14].The extracted ρ_C of both metals and metal-nitrides to crystalline doped-Sb_2Te is in the same range ($\rho_C \approx 10^{-7}\Omega.cm^2$). In the case of TiW and W electrode, ρ_C increases slightly when annealed at a high temperature. This could be attributed to diffusion or chemical reactions at the interface. In the crystalline state ρ_C is almost independent of the temperature which is a characteristic of tunneling.

CONCLUSIONS

In this work we present the contact resistance of doped-Sb_2Te in the amorphous and crystalline state with different CMOS compatible electrode materials W, TiW, Ta, TaN and TiN. From these measurements ρ_C is extracted and it is compared for these different electrode materials. In the amorphous state of doped-Sb_2Te, the ρ_C extracted for metal nitride electrodes ($\rho_C \approx 10^{-4}\Omega.cm^2$) is low compared to the corresponding metal electrodes ($\rho_c \approx 10^{-3}\Omega.cm^2$). In the crystalline state the extracted ρ_C is in the same range ($\rho_C \approx 10^{-7}\Omega.cm^2$) for all the electrode materials. The electrical properties of the metal-PCM contacts in both states show close similarities with metal to silicon contacts. Based on these measurements it is concluded that the charge transport through the metal to PCM interface is dominated by thermionic-field emission in the amorphous state and tunneling in the crystalline state. Knowledge of ρ_C is useful in the selection of the metal electrode for integration of phase change memory cells.

ACKNOWLEDGMENTS

This research was carried out under the project number MC3.07298 in the framework of the Research Program of the Materials innovation institute M2i (www.m2i.nl).

REFERENCES

1. K. Attenborough, G. Hurkx, R. Delhougne, J. Perez, M. Wang, T. Ong, L. Tran, D. Roy, D. Gravesteijn, and M. van Duuren, IEDM Tech. Dig., 648-651(2010).
2. D. Roy, M. A.A. in't Zandt, and Rob A.M. Wolters, IEEE Electron Device Lett., 31 (11), 1293 - 1295 (2010).
3. S.J Proctor and L.W Linholm, IEEE Electron Device Lett., 3 (10), 294-296 (1982).
4. N. Stavitski, J. H. Klootwijk, H. W. Van Zeijl, A. Y. Kovalgin and R. A. M. Wolters, IEEE Trans. Semiconductor manufacturing, 22 (1), 146-152 (2009).
5. A. Y. C. Yu, Solid-state Elect. 13, 239-247(1970)
6. F. A. Padovani and R. Saratton, Solid-state Elect. 9, 695-707(1966)
7. D. K. Schroder, Semiconductor material and device characterization, 3nd Edition, Wiley-Interscience (2006) ch. 3
8. R. Yokota, Jap. J. Appl. Phys., 28(8) 1407-1411(1989)
9. S. W. Ryu, J. Ho Lee, Y. B. Ahn, Choon H. Kim, B. S. Yang,G. H. Kim, S. G. Kim, Se-Ho Lee, C. S. Hwang, and H. J. Kim, Appl. Phys. Lett., 95, 112110 (2009)
10. J. C. Male, Brit. J. Appl. Phys., 18, 1543-1549 (1967)

11. S. Raoux and M. Wuttig, Phase change materials: Science and application, Springer Verlag (2009) ch 9
12. D. Ielmini, Y. Zhang, J. Appl. Phys. 102, 054517 (2007)
13. C. B. Thomas, J. Phys. D: Appl. Phys., 9, 2587-2596 (1976)
14. S.M Sze, Physics of semiconductor devices, 2nd Edition, Wiley-Interscience, New York (1981) ch. 5.

Mater. Res. Soc. Symp. Proc. Vol. 1337 © 2011 Materials Research Society
DOI: 10.1557/opl.2011.979

Investigation on Phase Change Behaviors of Si-Sb-Te Alloy: The Effect of Tellurium Segregation

Xilin Zhou[1, 2], Liangcai Wu[1], Zhitang Song[1], Feng Rao[1], Kun Ren[1, 2], Yan Cheng[1], Bo Liu[1], Dongning Yao[1], Songlin Feng[1], and Bomy Chen[3]
[1]State Key Laboratory of Functional Materials for Informatics, Laboratory of Nanotechnology, Shanghai Institute of Micro-system and Information Technology (SIMIT), Chinese Academy of Sciences (CAS), Shanghai 200050, P. R. China
[2]Graduate University of the Chinese Academy of Sciences, Beijing, 100049, P. R. China
[3]Silicon Storage Technology, Inc., 1171 Sonora Court, Sunnyvale, CA 94086, U.S.A.

ABSTRACT

In this study, novel $Si_2Sb_2Te_6$ phase change material is investigated in detail for the phase change memory application using transmission electron microscopy and X-ray photoelectron spectroscopy. The phenomenon that Te diffuses to the film surface during phase switching and successively evaporates out has been confirmed. The phase change memory cells employing $Si_2Sb_2Te_6$ and $Si_3Sb_2Te_3$ materials are fabricated and programmed. For the $Si_2Sb_2Te_6$-based cell a data endurance of 5×10^5 cycles is achieved with a failure mode resembling reset stuck, which can be attributed to the migration of Tellurium during the operation cycles. It means that a thermally stable material system of $Si_xSb_2Te_3$ is preferred for the PCM applications.

INTRODUCTION

Nowadays advanced applications from digital cameras to smart phones are driving the demand for novel nonvolatile memory (NVM) technology capabilities. Phase change memory (PCM) technology directly addresses the needs of current electronic systems with innovative key technology features, such as nonvolatility, high scalability, low power consumption, and good data retention [1]. Conventional phase change materials used in PCM are chalcogenide materials, especially the Te-based alloys. These alloys possess notable difference in electrical properties between the amorphous (reset state with high resistivity) and crystalline (set state with low resistivity) phases, which can be employed as bistable states of PCM in nonvolatile electronic storage.

Over the past few years, various material families have been found qualified for PCM application. Among these, Si-Sb-Te (SST) materials have been reported with low energy cost and good data detention due to its higher resistance and enhanced thermal stability compared with Ge-Sb-Te (GST) alloys [2–4]. Similar to the $Ge_2Sb_2Te_5$ ($GeTe$-Sb_2Te_3) system, the imitator $Si_2Sb_2Te_6$ pseudobinay alloy has been proposed along the Si_2Te_3-Sb_2Te_3 tie line [3]. The phenomenon of Te phase separation, however, is still observed in the Te-rich Si-Sb-Te materials ($Si_2Sb_2Te_{5-6}$) [3, 5], since Te in GST is confirmed to be diffused towards grain boundaries and interfaces at high annealing temperature [6, 7].Te segregation, which mainly ascribes to its low melting temperature and high vapor pressure, will lead stoichiometric deviation and deteriorate the reproducibility of the phase change cycles. In this study the phase separation evidence of a phase change material $Si_2Sb_2Te_6$ practiced for NVM applications is presented by *in situ* transmission electron microscopy (TEM) and X-ray photoelectron spectroscopy (XPS) investigation, and the effect of Te separation on the switching properties and long-term data endurance of PCM cells has been suggested.

EXPERIMENTAL DETAILS

The SST thin films are deposited at room temperature by sputtering the $Si_2Sb_2Te_6$ pure target. The background pressure in the sputtering system is less than 2.0×10^{-4} Pa. The compositions of the films are determined by means of energy-dispersive spectroscopy (EDS) in a scanning electron microscopy, which are obtained from the samples deposited on Al foils to void the disturbance of element from the Si/SiO_2 substrate. The microstructures of the $Si_2Sb_2Te_6$ films are analyzed through plan-view TEM (JEOL 2010) with an operating voltage of 200 kV. The bright field (BF) TEM images and selected area electron diffraction (SAED) patterns are obtained for the 40-nm thick $Si_2Sb_2Te_6$ films deposited on supporting grids coated with carbon film. The chemical bonding characteristics of the annealed $Si_2Sb_2Te_6$ films (~240 nm) on Si substrate is examined by XPS. The surface of the films are cleaned by Ar^+ sputtering before XPS experiments, and the instrumental resolution is estimated to be 0.1–1 eV. A 0.18 μm complementary metal-oxide-semiconductor (CMOS) technology is utilized to fabricate the PCM cells with a typical T-shaped structure, where the chalcogenide layer is sandwiched in between the top electrode and heater plug. The W heater plug with a diameter of 260 nm is covered by 100-nm thick crystalline $Si_2Sb_2Te_6$ film. Reactive ion etching process is used to pattern the SST layer, followed by the deposition of TiN (20 nm) and Al (300 nm) films which are then patterned to form the top electrode. The electrical switching (set/reset) and endurance characteristics of the PCM cells are monitored using an Agilent 81104A pulse generator and a Keithley-2400 meter.

RESULTS AND DISCUSSION

The typical plan-view TEM images of $Si_2Sb_2Te_6$ film *in situ* annealed during observation are presented in figure 1. A typical worm image and the halo diffraction patterns indicate the amorphous structure of the as-deposited $Si_2Sb_2Te_6$ film shown in figure 1(a). EDS attached to TEM is used to characterize the composition of a 5 μm^2 matrix, which determines the initial atomic concentration to be 18.81%, 21.90%, and 59.29% for Si, Sb, and Te element, respectively. As the annealing temperature increases, the matrix is interspersed with tens of dark points corresponding to the crystal grains of heavy elements (Sb or Te) in BF mode, as shown in figure 1(b). The black and white contrast, that is mass contrast, arises from incoherent elastic scattering (Rutherford scattering) of electrons: higher-mass areas (Sb and Te) will scatter more electrons off axis, thus appear darker than lower-mass areas (Si) for the case of a BF image. The diffraction rings in the inset of figure 1(b) are observed and identified as rhombohedral Sb_2Te_3 phase, indicating a polycrystalline structure. It is revealed in figure 1(c) that a few of the dark points evolve into nanowire-like pieces with an atomic ration Si 17.90%:Sb 9.26%:Te 72.84%. The segregation of Te-rich pieces at 483 K is also confirmed by the SAED pattern in the inset of figure 1(c), where the hexagonal (100) and (012) planes associated with the Te crystal emerge. As the annealing temperature reaches 573 K given in figure 1(d), these dark points or pieces totally disappear, instead, the annealed matrix presents a kind of phase morphology with even distributed bright and dark areas, and the composition of the film is close to $Si_{1.42}Sb_2Te_{2.95}$. A sole Sb_2Te_3 phase is also identified from the diffraction rings in the inset of figure 1(d).

The EDS results of the same 5 μm^2 area are recorded *in situ* throughout the whole process of

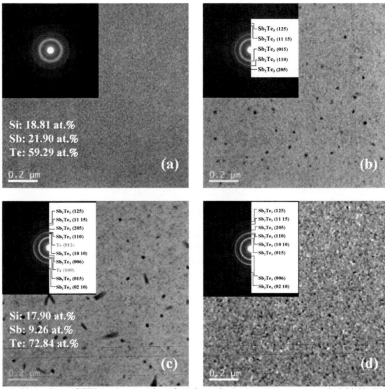

Figure 1. Plan-view of TEM micrographs of as-deposited (a) and *in situ* annealed Si₂Sb₂Te₆ film: (b) 423 K, (c) 483 K, and (d) 573 K. Insets show the SAED patterns of the corresponding matrix.

annealing. One can find a consistent tendency of Te decrease with respect to Si, Sb increase, as shown in figure 2. A final composition of ~ $Si_{1.50}Sb_2Te_{2.95}$ is obtained after annealed at 573 K for 10 min, as denoted by the open symbols in figure 2. Combining the results given in figure 1, it is indicate that a high temperature causes Te segregating to interfaces and a consequent change in stoichiometry to $Si_2Sb_{2-y}Te_{6-x}$ (albeit not as severe, there is also some motion of Sb [6], see below). It is rational that Te atoms are likely to bond with Sb atoms into stable Sb_2Te_3 phase in the final compound, leaving almost all Si atoms in the amorphous phase to form thermally stable $Si_xSb_2Te_3$ composition. It is worth to note that the desired phase of Si_2Te_3 doesn't appear during the annealing process. Since the higher crystallization temperature of Si_2Te_3 (>523 K [4]) than Sb_2Te_3, before the formation of Si_2Te_3 crystal, most of the Te atoms are prone to either separate from the $Si_2Sb_2Te_6$ film and successively evaporate out or bind in Sb_2Te_3 phase. Hence, it is suggest that a thermally stable Si_2Te_3-Sb_2Te_3 combined mode seems infeasible, and the phase segregation of the $Si_2Sb_2Te_6$ alloy at high temperature is unavoidable.

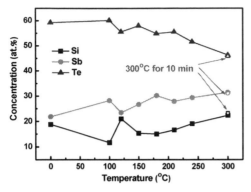

Figure 2. EDS results of as-deposited and *in situ* annealed $Si_2Sb_2Te_6$ film. The open symbols provide the corresponding atomic concentration after annealed at 573 K for 10 min.

XPS spectra of Sb 3d and Te 3d in the crystalline $Si_2Sb_2Te_6$ films (Ar^+ etching time 240 s) annealed at various temperatures are shown in figure 3(a) and 3(b). The Sb $3d_{3/2}$ of Sb_2O_5 and Sb metallic bonding are at 539.8 and 537–538 eV, while the Sb $3d_{5/2}$ of Sb_2O_5 and Sb metallic bonding are at 530.4 and 529–530 eV, respectively [8, 9]. The Sb homopolar (Sb–Sb) bonding is at 537.4 eV for Sb $3d_{3/2}$ and 527.9 or 528.2 eV for Sb $3d_{5/2}$, respectively [8, 9]. In figure 3 (a) the Sb peaks shift to the metallic bonding of Sb–Te as the annealing temperature increases, since Sb won't bond with Si [9], and the intensity decreases gradually. Relatively, the peak intensity of Sb oxide increases gradually. This indicates the mobility of Sb in the film to strength the oxidized Sb bonding first and the Sb–Te bonding remains. As the temperature goes up, however, the numbers of Sb–Te bonding decrease and their peak intensities are reduced, which should be attributed to the segregation of Te at high temperature. Te $3d_{3/2}$ and $3d_{5/2}$ oxide bonding states are at 576–577 and 586–587 eV, while the peaks located at 572.5–574 and 583–584 eV are related to metallic Te $3d_{5/2}$ and $3d_{3/2}$ bonding states, respectively [8, 9]. The Te homopolar peak (Te–Te) is at 573.1 and 583 eV for Te $3d_{5/2}$ and $3d_{3/2}$, respectively [8, 9]. As the annealing temperature

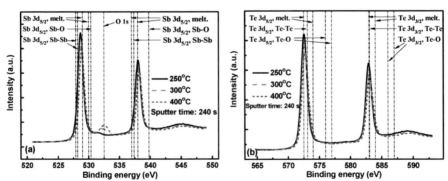

Figure 3. XPS spectra of (a) Sb 3d and (b) Te 3d annealed at various temperatures.

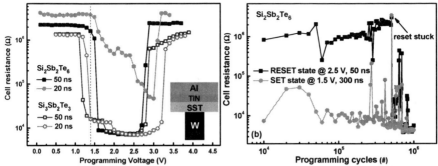

Figure 4. (a) Resistance–voltage characteristics of PCM cell based on $Si_2Sb_2Te_6$ and $Si_3Sb_2Te_3$ materials for different voltage pulse width, the inset is the schematic cross-sectional structure of the T-shaped PCM cell [3]; (b) endurance characteristic of the PCM cell base on $Si_2Sb_2Te_6$ material with a data endurance of 5×10^5cycles.

increases, the peak intensity decreases, as shown in figure 3(b). And the peak positions of Te $3d_{3/2}$ and $3d_{5/2}$ shift from metallic bonding to homopolar bonding, which means the broken of Sb–Te bonds as well as the formation of Te–Te bonds, thus further suggests the unavoidable Te motion to the interface since the oxidized Te bonding keeps almost constant.

PCM units with T-shaped structure using the $Si_2Sb_2Te_6$ and $Si_3Sb_2Te_3$ films are fabricated to evaluate the effects of Te separation of Te-rich SST films on the device performance. As mentioned above, the significant Te loss leaves a $Si_xSb_2Te_3$ film behind, so it is rational to compare the electrical phase change characteristics of the two phase change materials. Figure 4(a) provides a comparison of set/reset operations between the cells based on $Si_2Sb_2Te_6$ and $Si_3Sb_2Te_3$ for the set pulse widths of 20 and 50 ns. It is clear that the set voltage decreases with the pulse width for each cell, since the wider pulse thus longer heating period helps to decrease the required driving voltage [10]. The saturated reset voltages required for both of the cells are almost the same under corresponding pulse width, while for the $Si_3Sb_2Te_3$-based cell the voltage needed to set the active region is smaller as marked by the dash line in the figure. The sensing margin (defined as R_{reset}/R_{set}) for the cells is ~100, except the $Si_2Sb_2Te_6$ case programmed with 20 ns pulse, in which an incomplete set programming takes place. As is reported that increasing Te will slow the crystallization process [11], the pulse duration of 20 ns is therefore insufficient to set the $Si_2Sb_2Te_6$-based cell and leads a drift set state with programming voltage. It is the inevitable Te segregation in the $Si_2Sb_2Te_6$ film that should be responsible for the incomplete crystallization of the PCM unit, which will worsen the set resistance uniformity and degrade the cell performance with operation cycles (see below). One can extend the set pulse width, say, 50 ns, to remove the incomplete set programming by trading off the operation speed.

The endurance performance of a cell based on $Si_2Sb_2Te_6$ material is shown in figure 4(b). The fluctuation of set resistance is observed during the operation cycles, and similar to the reset stuck, the endurance failure occurs at about 5×10^5cycles, followed by a serious resistance disorder. On the other hand, a larger sensing margin and more stable set state during cycling measurements are achieved in the $Si_3Sb_2Te_3$-based cell [4]. What's more, the data endurance survives for more than 10^7 cycles for the cell using the thermally stable Te-poor $Si_3Sb_2Te_3$ composition. Reasonably, the contrast in endurance property between these two materials can be ascribed to the Te motion in the Te-rich $Si_2Sb_2Te_6$ film with cycles, which can become a serious reliability issue during the storage operations of a PCM cell.

CONCLUSIONS

In summary, the phase change behaviors of the Te-rich $Si_2Sb_2Te_6$ film is investigated using TEM and XPS methods. It is found that the migration of Te to the grain boundaries and interfaces is unavoidable in the heating process, which leads the thermally unstable of the $Si_2Sb_2Te_6$ material, evolving to a stable composition of $\sim Si_{1.50}Sb_2Te_{2.95}$. The electrical switching properties of PCM cells using $Si_2Sb_2Te_6$ and $Si_3Sb_2Te_3$ films are presented and compared. The phenomenon of Te separation from the $Si_2Sb_2Te_6$ alloy should be responsible for the deteriorated reproducibility of the phase change cycling process. We conclude that the optimized composition of the SST family should focus on the thermally stable $Si_xSb_2Te_3$ system so as to meet the demand for the commercial applications of PCMs.

ACKNOWLEDGMENTS

Financial supported by National Integrate Circuit Research Program of China (2009ZX02023-003), National Basic Research Program of China (2010CB934300, 2011CB309602, 2011CB932800), The author Liangcai Wu gratefully acknowledges the support of K. C. Wong Education Foundation, Hong Kong.

REFERENCES

1. M. Wuttig, *Nat. Mater.* **4**, 265, (2005).
2. T. Zhang, Z. T. Song, F. Rao, G. M. Feng, B. Liu, S. L. Feng, and B. Chen, *Jpn. J. App. Phy.* **46**, L247 (2007).
3. X. L. Zhou, L. C. Wu, Z. T. Song, F. Rao, B. Liu, D. N. Yao, W. J. Yin, J. T. Li, S. L. Feng, and B. Chen, *Appl. Phys. Express* **2**, 091401(2009).
4. F. Rao, Z. T. Song, K. Ren, X. L. Zhou, Y. Cheng, L. C. Wu, and B. Liu, *Nanotechnology* **22**, 145702 (2011).
5. Y. Cheng, X. D. Han, X. Q. Liu, K. Zheng, Z. Zhang, T. Zhang, Z. T. Song, B. Liu, and S. L. Feng, *Appl. Phys. Lett.* **93**, 183113 (2008).
6. C. Cabral, Jr., K. N. Chen, L. Krusin-Elbaum, and V. Deline, *Appl. Phys. Lett.* **90**, 051908 (2007).
7. L. Krusin-Elbaum, C. Cabral, Jr., K. N. Chen, M. Copel, D. W. Abraham, K. B. Reuter, S. M. Rossnagel, J. Bruley, V. R. Deline, *Appl. Phys. Lett.* **90**, 141902 (2007).
8. M. J. Shin, D. J. Choi, M. J. Kang, and S. Y. Choi, *J. Korea Phys. Soc.* **44**, 10 (2004).
9. H. Y. Cheng, C. A. Jong, R. J. Chung, T. S. Chin, and R. T. Huang, *Semicond. Sci. Technol.* **20**, 1111 (2005).
10. D. S. Chao, C. Lien, C. M. Lee, Y. C. Chen, J. T. Yeh, F. Chen, M. J. Chen, P. H. Yen, M. J. Kao, and M. J. Tsai, *Appl. Phys. Lett.* **92**, 062108 (2008).
11. K. Ren, F. Rao, Z. T. Song, Yan Cheng, L. C. Wu, X. L. Zhou, Y. F. Gong, M. J. Xiao, B. Liu, and S. L. Feng, *Scripta Mater.* **64**, 685 (2011).

Mater. Res. Soc. Symp. Proc. Vol. 1337 © 2011 Materials Research Society
DOI: 10.1557/opl.2011.977

Recent Progress in Downsizing FeFETs for Fe-NAND Application

Le Van Hai, Mitsue Takahashi, Shigeki Sakai
National Institute of Advanced Industrial Science and Technology
Central 2, 1-1-1 Umezono, Tsukuba, Ibaraki, 305-8568, Japan

ABSTRACT

Sub-micrometer ferroelectric-gate field-effect transistors (FeFETs) of 0.56 μm and 0.50 μm gate lengths were successfully fabricated for Fe-NAND cells. Gate stacks of the FeFETs were Pt/SrBi$_2$Ta$_2$O$_9$(SBT)/Hf-Al-O/Si. The gate stacks were formed by electron beam lithography and inductively coupled plasma reactive ion etching (ICP-RIE). Ti and SiO$_2$ hard masks were used for the 0.56 μm- and 0.50 μm-gate FeFETs, respectively, in the ICP-RIE process. Steep SBT sidewalls with the angle of 85° were obtained by using the SiO$_2$ hard masks while 76° sidewalls were shown using Ti hard masks. All fabricated FeFETs showed good electrical characteristics. Drain current hysteresis showed larger memory windows than 0.95 V when the gate voltages were swung between 1±5 V. The FeFETs showed stable endurance behaviors over 10^8 program/erase cycles. Drain current retention properties of the FeFETs were good so that the drain current on/off ratios did not show practical changes after 3 days.

INTRODUCTION

Nowadays, flash memories have seen a high growth market, driven by the demand for various portable electronic appliances, like mobile phones and digital cameras. Floating-gate type (FG) NAND flash memories have played a key role and widely used in the products. However, the FG NAND will be no longer most suitable for high-performance non-volatile memory products like enterprise solid state drives which require low-power consumption and high reliability [1]. Ferroelectric-gate field-effect transistors (FeFETs) have attracted much attention as nonvolatile memory transistors [2-21]. Since a long data retention of an FeFET with Pt/SrBi$_2$Ta$_2$O$_9$(SBT) /Hf-Al-O(HAO)/Si gate was achieved for the first time [5], we have made much progress in the FeFET fabrication process [5-10, 14,15,17-21]. We also have intensively studied the FeFET application to NAND (Fe-NAND) flash memories [6,19-21] because the FeFETs showed good characteristics such as long retentions, low operation voltages of 6 V and high endurance of 10^8 cycles for program/erase (P/E). The operation voltage is much lower and the endurance is much higher than those of the conventional FG NAND, which has 20V program voltage and 10^4 P/E cycles. Furthermore, the Fe-NAND flash memory has a good potential for scaling toward sub-10 nm, whereas the FG NAND flash memory is supposed to have intrinsic difficulty in downsizing to that scale [6].

In this study, we fabricated the Pt(250nm)/SBT(200nm)/HAO(7nm)/Si FeFETs and investigated their downsizing with keeping their good electrical properties. Electron-beam (EB) lithography and inductively coupled plasma reactive ion etching (ICP-RIE) using hard masks were key processes. We prepared two kinds of hard masks. One was a metal Ti hard mask made by a lift-off technique, the other was an insulator SiO$_2$ hard mask by the ICP-RIE. A 0.56 μm FeFET with the SBT sidewall angle of 76° was fabricated by using the Ti hard mask. By using the SiO$_2$ hard mask and by optimizing the ICP-RIE condition, a 0.50 μm FeFET with 85° SBT sidewall was successful fabricated.

EXPERIMENT

A schematic cross section of a completed device is shown in Fig. 1. The common fabrication process for FeFETs with gate lengths (Ls) of L=0.56μm and L=0.50μm was schematically shown in Fig. 2. A p-type Si substrate with deeply doped n$^+$ regions for the source/drain contact was prepared beforehand. First, a sacrificial SiO$_2$ layer of substrate was removed by buffered hydrofluoric acid (BHF) solution. Then, a stack of Pt/SBT/HAO film was deposited by a large-area-type pulse-laser-deposition [22] and EB evaporator. The deposition steps mentioned above were described in detail in our previously reported [5-10, 14,15,18].

In technique 1, The Ti mask was fabricated by a lift-off technique. First a positive EB resist (polymethyl methacrylate) layer was prepared by a spin coating on the Pt/SBT/HAO/Si stack. Then the 0.3 μm gate patterns were formed by an EB lithography system. After the gate patterns were developed, a 300 nm-thick Ti was deposited over the patterned EB resist. Finally the EB resist was completely lifted off the sample in acetone, and the Ti hard mask pattern was left on the Pt. Figure 3 (a) showed a cross-sectional scanning electron microscope (SEM) image of the Ti mask. The ICP-RIE was used for etching the Pt/SBT/HAO stack with the Ti mask by the condition in Table 1.

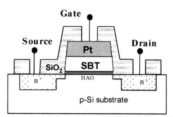

Figure 1. Cross-sectional drawing of Pt/SBT/HAO/Si FeFET gate.

Table 1. Typical parameters used in ICP-RIE processes.

	Ar flow (sccm)	Etchant gas/ flow (sccm)	Pressure (Pa)	Antenna RF power(W)	Bias RF power (W)
Technique 1	5	BCl$_3$/5	0.2	600	360
Technique 2-SBT, SiO$_2$	5	BCl$_3$/10	0.3	600	600
Technique 2- Pt	7	Cl$_2$/3	0.15	600	500

In technique 2, the SiO$_2$ mask was fabricated by a copy mask technique. First, a 300-nm-thick SiO$_2$ layer was deposited by sputtering on the Pt/SBT/HAO/Si stack. Next, a negative EB resist (SAL-601 SR7) layer was prepared by a spin coating over the SiO$_2$. Then the 0.3 μm gate patterns were formed by the EB lithography system. Figure 3 (b) showed an SEM image of the EB resist pattern after the development. The ICP-RIE system was used to transfer the EB resist patterns to the SiO$_2$ layer. Finally the SiO$_2$ hard mask was formed for following gate etching process. The SiO$_2$ hard mask was used for both Pt and SBT etching by the ICP-RIE by the condition listed in table 1.

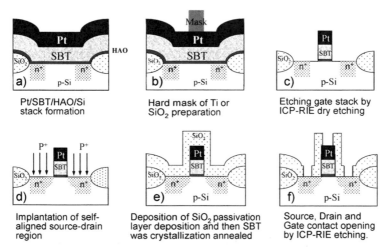

a) Pt/SBT/HAO/Si
stack formation

b) Hard mask of Ti or
SiO₂ preparation

c) Etching gate stack by
ICP-RIE dry etching

d) Implantation of self-
aligned source-drain
region

e) Deposition of SiO₂ passivation
layer deposition and then SBT
was crystallization annealed

f) Source, Drain and
Gate contact opening
by ICP-RIE etching.

Figure 2. Processing sequence from (a) to (f) for fabricating Pt/SBT/HAO/Si FeFET gates.

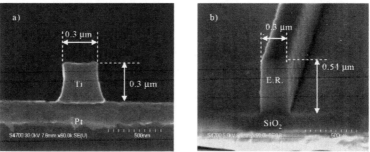

Figure 3. SEM cross-sectional images of (a) Ti hard mask, and (b) a negative EB resist pattern for SiO₂ hard mask.

After the gate etching process, phosphorus ions were implanted to form lightly doped source-and-drain regions by the self-aligned gate technique. A 200-nm SiO₂ passivation layer was deposited subsequently by rf magnetron sputtering in order to protect the SBT sidewalls and to isolate neighboring FeFETs. The substrate was annealed at 775 °C for 30 min in oxygen ambience for poly-crystallization of the SBT and activation of the implanted phosphorus. Contact holes of gate, source and drain were formed by the ICP-RIE. The electrical properties of FeFETs were investigated. Sidewall angles and cross sections of the FeFETs were observed by SEM as shown in Figs. 4 (a) and (b).

DISCUSSION

Figures 3 a) and b) showed cross-sectional images of the Ti and EB-resist mask profiles, respectively, by the EB scanning of the 0.3-μm long gate patterns. They showed sharp profiles with steep sidewall angles. The Ti thickness was 300 nm. The EB resist thickness was 540 nm.

Figures 4 a) and b) showed cross-sectional SEM images of the FeFET gates which were etched by the ICP-RIE using the Ti hard mask and the SiO_2 hard mask, respectively. SiO_2 passivation layers were deposited after the gate etchings. The SBT sidewall angles were 76° in figure a) and 85° in figure b). That means the technique 2 is more suitable than technique 1 for ongoing downscaling. Metallurgical gates of L=0.56 μm and L=0.50 μm were estimated from Figs. 3 a) and b), respectively.

Drain current-gate voltage (I_d-V_g) curves of the Pt/SBT/HAO/Si FeFETs were shown in Fig. 5(a) for L = 0.56 μm and in Fig. 5 (b) for L = 0.50 μm. They showed good counter-clockwise hysteresis loops when V_gs were swept in the ranges of 1±2 V, 1±3 V, 1±4 V and 1±5 V. Static I_d-V_g memory windows of 0.95 V were obtained for both the 0.56 μm- and 0.50 μm-FeFETs, when the V_g were scanned between -4 V and 6 V.

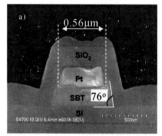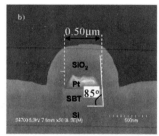

Figure 4.   SEM cross-sectional images of the FETs with (a) L = 0.56 μm was fabricated follow Ti hard mask technique and (b) L = 0.50 μm fabricated follow SiO_2 hard mask technique. The FeFETs were covered by a 200-nm-thick SiO_2 passivation layer.

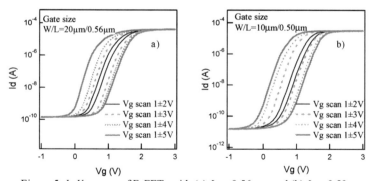

Figure 5. I_d-V_g curves of FeFETs with (a) L_m= 0.56 μm and (b) L_m= 0.50 μm.

Endurance performances of these FeFETs were investigated. As indicated in Figs. 6(a) and (b), I_d-V_g cures of the FeFETs with $L = 0.56$ μm and $L = 0.50$ μm were measured after multiple P/E cycle application. The input pulses were +6 V and -4 V with 20 μs period. Figures 7(a) and (b) showed their endurance properties. Both FeFETs showed good endurance performances with little threshold voltage (V_{th}) shift even after 10^8 P/E pulse cycles. The V_{th}s of both sides did not exhibit significant changes after the 10^8 pulse applications. The results indicated that the Fe-NAND has 10^4 times better endurance than the conventional FG-NAND.

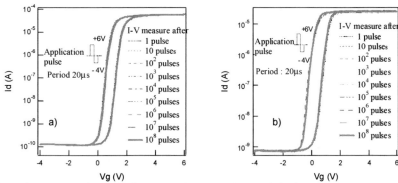

Figure 6. Demonstration of endurance performance characteristics by I_d-V_g cures measured after multiple P/E cycle application of FeFETs' (a) $L = 0.56$ μm and (b) $L = 0.50$ μm.

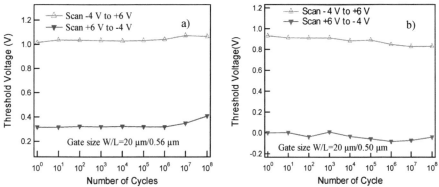

Figure 7. Endurance properties of the FeFETs with gate length (a) $L = 0.56$ μm and (b) $L = 0.50$ μm.

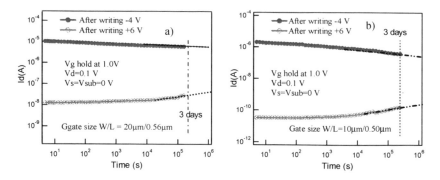

Figure 8. Drain current retention characteristics of FeFETs with gate length (a) L= 0.56 μm and (b) L = 0.50 μm.

Figures 8 (a) and (b) showed I_d retention characteristics of FeFETs with $L = 0.56$ μm and $L = 0.50$ μm, respectively. In these measurements, writing voltages of +6 V and -4 V with 0.1 s pulse widths were applied to the gate electrode, then I_d versus time characteristics were investigated with drain voltage (V_d), source voltage (V_s) and substrate voltage (V_{sub}) of $V_d = 0.1$ V, $V_s = 0$ V and $V_{sub} = 0$ V, respectively. The hold bias voltage (V_{hold}) of 1.0 V was applied to the gate during the retention measurements. As shown in Figs. 7 (a) and (b), the on/off I_d ratio did not significantly change for 3-day measurements. Both the $L = 0.50$ μm and $L = 0.56$ μm FeFETs showed good retention properties which indicated 10 year retention times by their extrapolated lines.

CONCLUSIONS

Pt/SBT/HAO/Si FeFETs with the metallurgical gate lengths of 0.56 μm and 0.50 μm were successfully fabricated and characterized. The Ti lift-off hard mask, SiO_2 copy hard mask and ICP-RIE techniques were employed for gate patterning processes to investigated the scalability of the FeFETs. Cross-sectional profiles of the 0.50 μm FeFET gates made by using the SiO_2 hard mask showed a steep sidewall angle of 85°. All FeFETs exhibited good electrical characteristics. Memory windows of the FeFETs were 0.95 V, when V_g scanned between 1±5 V. The FeFETs showed good endurance performances up to 10^8 cycles that is larger 10^4 times than the endurance of FG-NAND. Drain current retention properties of FeFETs showed good behaviors and drain current on/off ratio without practically change after 3 days.

ACKNOWLEDGMENTS

This work was supported by New Energy and Industrial Technology Development Organization (NEDO) and by AIST grants for the focused researches and for the leading innovations. We thank H. Nakai and A. Nitayama in Toshiba Corporation and K. Takeuchi in the University of Tokyo for valuable discussions.

REFERENCES

1. S. Lai, *IEEE IEDM Techn. Dig.*, pp.255-258 (2003).
2. Y. Tarui,T. Hirai, K. Teramoto, H. Koike and K. Nagashima, *Appl. Surf. Sci.* **113**, pp.656-663 (1997).
3. Y. Higuma, Y. Matsui, M. Okuyama, T. Nakagawa and Y. Hamakawa, *Jpn. J. Appl. Phys.* **17** Suppl. 17-1, pp.209-214 (1978).
4. J. F. Scott, *Ferroelectric Memories*, (Springer, 2000) chapter 12.
5. AIST press release, *"Development of the 1T FeRAM : Towards the Realization of the Ultra-Gbit Next-Generation Semiconductor Memory,"* Oct 24, 2002.
6. S. Sakai, M. Takahashi, K. Takeuchi, Q.-H. Li, T. Horiuchi, S. Wang, K.-Y. Yun, M. Takamiya and T. Sakurai, *Proc. of the 23rd IEEE Non-Volatile Semiconductor Memory Workshop and 3rd International Conf. on Memory Technology and Design*, pp 103-105 (2008).
7. S. Sakai, US patent 7,226,795 (2005).
8. S. Sakai and R. Ilangovan, *IEEE Electron. Devices Lett.* **25**, pp.369-371 (2004).
9. S. Sakai, M. Takahashi and R. Ilangovan, *IEEE IEDM Tech. Dig.*, pp.915-918 (2004).
10. S. Sakai, M. Takahashi and R. Ilangovan, *Jpn. J. Appl. Phys.* **43**, pp.7876-7878 (2004).
11. K. Aizawa, B.-E. Park, Y. Kawashima, K. Takahashi and H. Ishiwara, *Appl. Phys. Lett.* **85**, pp.3199-3201 (2004).
12. S.I. Shim, Y.S. Kwon, S.-I. Kim, Y.T. Kim and J.H. Park, *physica status solidi (a)* **201**, pp. R65–R68 (2004).
13. S.I. Shim, Y.S. Kwon, S.I. Kim, Y.T. Kim and J.H. Park, *J. Vac. Sci. Technol. A* **22**, pp.1559-1563 (2004).
14. M. Takahashi and S. Sakai, *Jpn. J. Appl. Phys.* **44**, L800-L802 (2005).
15. Q.-H. Li and S. Sakai, *Appl. Phys. Lett.*, **89**, 222910 (2006).
16. X.-B. Lu, K. Maruyama and H. Ishiwara, *Semicond. Sci. Technol.* **23**, 045002 (2008).
17. T. Horiuchi, M. Takahashi, Q.-H. Li, S. Wang and S. Sakai, *Semicond. Sci. Technol.* **25**, 055005 (2010).
18. L. V. Hai, M. Takahashi and S. Sakai, *Semicond. Sci. Technol.* **25**, 115013 (2010).
19. S. Wang, M. Takahashi, Q.-H Li, K. Takeuchi and S. Sakai, *Semicond. Sci. Technol.* **24**, 105029 (2009).
20. T. Hatanaka, R. Yajima, T. Horiuchi, S. Wang, X. Zhang, M. Takahashi, S. Sakai and K. Takeuchi, *IEEE Symp. VLSI Circuits Dig. of Tech. Papers*, pp.78-79 (2009).
21. S. Sakai and M. Takahashi, *Materials* **3**, pp.4950-4964 (2010).
22. S. Sakai, M. Takahashi, K. Motohashi, Y. Yamaguchi,N. Yui and T. Kobayashi, *J. Vac. Sci. Technol. A* **25**, pp.903-907 (2007).

Mater. Res. Soc. Symp. Proc. Vol. 1337 © 2011 Materials Research Society
DOI: 10.1557/opl.2011.1029

Lanthanum Oxide Capping Layer for Solution-Processed Ferroelectric-Gate Thin-Film Transistors

Tue T. Phan[1], Trinh N. Q. Bui[2], Takaaki Miyasako[2], Thanh V. Pham[1], Eisuke Tokumitsu[2,3], and Tatsuya Shimoda[1,2]

[1]School of Materials Science, Japan Advanced Institute of Science and Technology, 1-1 Asahidai, Nomi, Ishikawa 923-1292, Japan.
[2]Japan Science and Technology Agency, ERATO, Shimoda Nano-Liquid Process Project, 2-5-3 Asahidai, Nomi, Ishikawa 923-1211, Japan.
[3]Precision and Intelligence Laboratory, Tokyo Institute of Technology, 4259-R2-19 Nagatsuta, Midori-ku, Yokohama 226-8503, Japan.

ABSTRACT

We report on the use of La_2O_3 (LO) as a capping layer for ferroelectric-gate thin-film transistors (FGTs) with solution-processed indium-tin-oxide (ITO) channel and $Pb(Zr,Ti)O_3$ (PZT) gate insulator. The fabricated FGT exhibited excellent performance with a high "ON/OFF" current ratio (I_{ON}/I_{OFF}) and a large memory window (ΔV_{th}) of about 10^8 and 3.5 V, respectively. Additionally, a significantly improved data retention time (more than 16 hours) as compared to the ITO/PZT structure was also obtained as a result of good interface properties between the ITO channel and LO/PZT stacked gate insulator. We suggest that the LO capping layer acts as a barrier to prevent the interdiffusion and provides atomically flat ITO/LO/PZT interface. This all-oxide FGT device is very promising for future ferroelectric memories.

INTRODUCTION

All-oxide ferroelectric-gate thin-film transistors (FGTs) are very promising and have drawn much attention for ultimate system-on-panel or system-on-film applications [1-11]. In particular, the use of indium-tin-oxide (ITO) as a channel layer and ferroelectric $(Bi,La)_4Ti_3O_{12}$ (BLT) or $Pb(Zr,Ti)O_3$ (PZT) film as a gate insulator has been demonstrated as promising candidates for these targets [6-9,11]. Generally, the use of PZT is preferred to BLT because of its advantages, such as higher remanent polarization, lower crystallization temperature and more flat surface. However, component interdiffusion between solution-processed ITO and PZT layers severely occurs resulting in a degradation of both ITO crystalline quality and ITO/PZT interface properties. As a result, most of these FGT memories exhibit very poor retention property, which is the main obstacle for practical use of it. Thus, preventing the interdiffusion is crucial to realize a good interface property which must be indispensable for obtaining good retention characteristics.

In the case of Si-based metal-ferroelectric-semiconductor (MFS) memory, an insulator (I) buffer layer is often used to form the MFIS structure so that the interdiffusion or reaction between the ferroelectrics and Si substrate is hindered [12,13]. For our FGT structure, a capping layer can also be considered to prevent the interdiffusion.

Lanthanum oxide (LO), a well-known high-κ dielectric material, has a tetragonal structure with a lattice constant (3.937-4.05 Å) close to that of PZT (4.03 Å) [14]. It is also thermodynamically stable without forming an interface layer when placed in contact with PZT

[15]. Surface morphology of LO thin layer is known to be very smooth [16]. The other key advantage of LO is the high band offset barrier, which reduces the conduction current of electrons and holes [14]. LO also has a strong electrical break-down field. Based on these considerations, in this study we made an effective approach to stabilize PZT surface and ITO/PZT interface in order to improve the retention property of the device: using a thin LO layer as a capping layer between ITO channel and ferroelectric PZT. To our knowledge, FGT with the ITO/LO/PZT heterostructure has not been reported by other researchers.

EXPERIMENTAL PROCEDURE

Bottom-gate structure TFTs were fabricated using ITO as a channel and PZT as a gate insulator. First, Pt/Ti (100 nm/10 nm) was deposited on an insulating substrate by sputtering and patterned as a bottom-gate electrode. Then, PZT gate insulator (225 nm) was formed by a sol-gel technique using an alkoxide-based PZT (Pb/Zr/Ti = 120/40/60) precursor solution (Mitsubishi Chemical Co.). This solution was multiply spin-coated, dried at 240 °C in air for 5 min and consolidated at 400 °C for 10 min. Then, a gate contact hole was formed by photolithography using wet-etching. After that, the PZT film was crystallized at 600 °C for 20 min in ambient air. As the thin capping layer, a LO layer was then spin-coated, slowly heated up to 550 °C (10 °C/min), and held for 10 min in O_2. In the next fabrication step, Pt source and drain electrodes were sputtered at RT and patterned by a lift-off process. Finally, an ITO layer was deposited by spin-coating using a carboxylate-based precursor solution (5 wt % SnO_2-doped), followed by annealing at 450 °C for 30 min in air. The channel length and width are 30 and 60 μm, respectively. Schematic cross-sections of the fabricated devices with and without capping layer are shown in Figure 1.

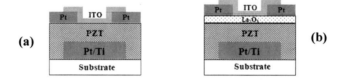

(a)   (b)

Figure 1. (a) Schematic cross-sectional view of ITO channel FGTs without capping layer and (b) with the La_2O_3 capping layer.

RESULTS AND DISCUSSION

Cross-sectional dark-field TEM images of the samples with and without the capping layer are displayed in Figure 2(a1) and 2(b1), respectively. Sharp ITO/LO/PZT heterointerfaces were observed for the structure with LO capping layer but not for the ITO/PZT structure. Fig. 2(a2) and 2(b2) correspond to high-resolution images focused on the ITO/PZT and ITO/LO/PZT interfaces, respectively. The high resolution image at ITO/PZT interface (Figure 2(a2)) exhibited an amorphous layer having a thickness of 7 to 10 nm between ITO and PZT even at as low as 450 °C annealing. In addition, energy dispersive X-ray spectrometry (EDX) analysis revealed that whereas Zr was segregated at the PZT surface, about 10-20 % of Pb and Ti diffused into ITO layer. These mean that there was some interaction between ITO and PZT after the ITO/PZT structure was annealed at 450 °C. In contrast, well-defined ITO/LO/PZT interfaces were observed with no reaction/defective layer (Figure 2(b2)). Thus, the LO layer served as a good barrier layer to suppress the element interdiffusion and reaction between the ITO and the PZT layers.

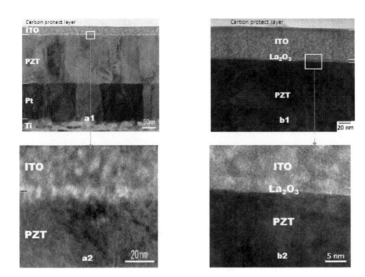

Figure 2. Cross-section TEM images of (a1) ITO/PZT/Pt and (b1) ITO/LO/PZT/Pt structures and HRTEM images focused on (a2) ITO/PZT and (b2) ITO/LO/PZT interfaces.

Figure 3(a) shows the 1 kHz capacitance-voltage (C-V) characteristics of the Pt/PZT/Pt (MFM) and Pt/LO/PZT/Pt (MFIM) capacitors. Our 225 nm PZT layer with the LO capping layer (MFIM) displayed the capacitance value of 0.92-2.73 $\mu F/cm^2$ and the double coercive voltage $2V_c = 3.84$ V, while the PZT layer without a capping layer showed 0.96-3.48 $\mu F/cm^2$ and the double coercive voltage $2V_c = 1.81$ V. Also, the dielectric constant of LO was estimated to be 26. A lower capacitance observed from the MFIM structure is attributed to the serial connection of two capacitors. The increase of coercive voltage of the MFIM structure was also observed from polarization-electrical field characteristics (Figure 3(b)). This is expected to result in an expansion of the device's memory window due to the relation: $2t_{ITO}E_c = \Delta V_{th}$, where the memory window is defined as threshold voltage shift ΔV_{th}. In addition, remanent polarizations of 41 and 52 $\mu C/cm^2$ were obtained for MFM and MFIM structures, respectively, which are typical values for PZT film.

Figure 4(a) shows drain current-gate voltage (I_D-V_G) characteristics of the FGTs with a channel length and width of 30 and 60 μm, respectively. The I_D-V_G curves were measured between ±7 V of scan voltage while V_D was kept at 1.5 V. We observed counter-clockwise hysteresis loops due to the nature of ferroelectric polarization as indicated by arrows, which confirm the nonvolatile memory function of the devices.

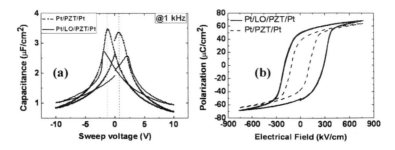

Figure 3. (a) *C-V* curves obtained at 1 kHz and (b) *P-E* hysteresis loops of two types of capacitor: Pt/PZT/Pt (dash line) and Pt/LO/PZT/Pt (solid line).

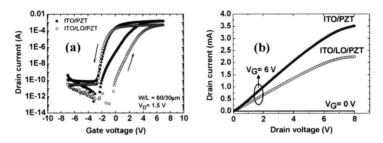

Figure 4. (a) I_D-V_G and (b) I_D-V_D characteristics of the ITO/PZT and ITO/LO/PZT FGTs.

In particular, the ITO/PZT structure, a maximum I_{ON}/I_{OFF} ratio of higher than 10^6 and a memory window of 1.0 V were obtained (Figure 4(a), solid circle). However, the difference in the drain currents between the binary states at a zero gate voltage, which is indispensable for nonvolatile memories, is not sufficiently large owing to a shift in the threshold voltage from "*OFF*" to "*ON*" state (V_{th}^+) to the negative voltage side. Such a shift in the threshold voltage suggests the existence of space charges in the ferroelectric PZT film or at the ITO/PZT interface similarly observed in Ref. [10,17,18], which can be anticipated from the HRTEM observation above. In addition, V_{th}^+ was also instable as I_D-V_G measurement repeated several times, which may be related to a charge injection into the interlayer under high electrical field. On the other hand, the new ITO/LO/PZT structure showed a large modulation of drain current between the accumulation and the depletion (10^8). Reasonably large memory window 3.5 V was obtained (Figure 4(a), open circle). These properties are much better than those of most of previously reported FGTs [1-11]. Furthermore, V_{th}^+ was very close to a zero bias indicating that the amount of space charge in the PZT film and at the ITO/LO/PZT interfaces is relatively low [5,17,18]. Figure 4(b) shows the drain current-drain voltage (I_D-V_D) characteristics obtained from the two structures. The maximum I_D of the FGT without the capping layer was around 3.6 mA under a bias condition of a $V_D = 8$ V ($V_G = 6$ V), while that of the other FGT with the capping layer was about 2.4 mA, which reflects the results from the transfer curves in Figure 4(a).

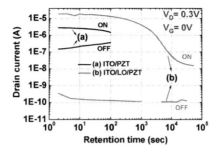

Figure 5. Retention properties of (a) ITO/PZT and (b) ITO/LO/PZT FGTs measured at RT.

In contrast to the result of reduced "*ON*" current (I_{ON}), the retention property of the device with the LO capping layer was much improved compared to that of the device without the capping, as seen in Figure 5. We first applied the gate voltage of +7 V and -7 V to write the data and the gate voltages were held at 0 V for the two devices. It was found that "*OFF*" current (I_{OFF}) of the ITO/PZT structure was very large and rapidly increased with retention time, whereas those of the new ITO/LO/PZT structure was relatively small. Moreover, the I_{ON} change of the ITO/LO/PZT structure was not so serious. After 16 hours, the I_{ON}/I_{OFF} ratio of the new ITO/LO/PZT structure was still about 10^2. These result demonstrated that the data retention time was dramatically improved by inserting the LO capping layer.

It has been widely accepted that the short retention time of the ferroelectric-gate field-effect transistors are mainly due to depolarization field and finite gate leakage current [19,20]. In our conventional ITO/PZT structure, the interfacial layer probably has low dielectric constant and stores a large amount of trapped-charges shielding the polarization, and hence reducing the retention time due to a rise of depolarization field. In addition, the 20-nm-surface layer may lose a perovskite structure and ferroelectricity upon the severe out-diffusion of Pb and Ti. These must alter the way of the ferroelectric polarization to control the semiconductor channel conductance. In contrast, the interdiffusion and formation of the interlayer were effectively prevented by the new ITO/LO/PZT structure. Thus, the depolarization field could be diminished, which in turn enhances the retention property. Moreover, the leakage current of the new ITO/LO/PZT structure was found to be one order of magnitude smaller than that of the conventional ITO/PZT structure due to higher band-offsets for carrier injection at ITO/LO/PZT interfaces (data not shown). Low leakage current could be considered as another reason for the improved retention property. However, the retention characteristic is still poor, which is probably due to non-uniformity of ferroelectric PZT polarization and low ITO crystallinity. Hence, improvements of these properties will be our next target.

CONCLUSIONS

The LO capping layer was proposed to integrate into the ITO/PZT FGT structure. The fabricated ITO/LO/PZT FGT device exhibited excellent performance with the high I_{ON}/I_{OFF} ratio and the large memory window of 10^8 and 3.5 V, respectively. Furthermore, an improved data retention time was also experimentally demonstrated. These improvements are associated with the good interface properties between ITO channel and stacked LO/PZT gate insulator. We also showed that the LO layer not only prevent the interdiffusion but also stabilize the PZT surface structure.

ACKNOWLEDGMENTS

This work was partially supported by Japan Science and Technology Agency-ERATO-Shimoda Nano Liquid Process Project. The authors would like to thank Mr. Higashimine (CNMT, JAIST) and Mr. Fujita (Kobelco Research Institute, Japan) for HRTEM observation.

REFERENCES

1. S. Mathews, R. Ramesh, T. Venkatesan, J. Benedetto, *Science* **276**, 238 (1997).
2. M. W. J. Prins, S. E. Zinnemers, J. F. M. Cillessen, and J. B. Giesbers, *Appl. Phys. Lett.* **70**, 458 (1997).
3. G. Hirooka, M. Noda, and M. Okuyama, *Jpn. J. Appl. Phys.* **43**, 2190 (2004).
4. T. Fukushima, T. Yoshimura, K. Masuko, K. Maeda, A. Ashida, and N. Fujimura, *Jpn. J. Appl. Phys.* **47**, 8874 (2008).
5. Y. Kato, Y. Kaneko, H. Tanaka and Y. Shimada, *Jpn. J. Appl. Phys.* **47**, 2719(2008).
6. T. Miyasako, M. Senoo, and E. Tokumitsu, *Appl. Phys. Lett.* **86,** 162902 (2005).
7. E. Tokumitsu, M. Senoo, and T. Miyasako, *Microelectronic Engineering* **80**, 305 (2005).
8. P. T. Tue, T. Miyasako, B. N. Q. Trinh, E. Tokumitsu, and T. Shimoda, *Ferroelectrics* **405**, 281 (2010).
9. T. Miyasako, B. N. Q. Trinh, M. Onoue, T. Kaneda, P. T. Tue, E. Tokumitsu, and T. Shimoda, *Appl. Phys. Lett.* **97**, 173509 (2010).
10. E. Tokumitsu and T. Oiwa, *Mater. Res. Soc. Symp. Proc.* **1250**, G13-07 (2010).
11. P. T. Tue, B. N. Q. Trinh, T. Miyasako, E. Tokumitsu, and T. Shimoda, *Microelectronics, 2010 IEEE International Conference on*, pp.32-35, 19-22 Dec. 2010.
12. S. Sakai, R. Ilangovan, IEEE Electron Dev. Lett. **25**, 369 (2004).
13. G. Hirooka, M. Noda, and M. Okuyama, *Jpn. J. Appl. Phys.* **43**, 2190 (2004).
14. G. D. Wilk, R. M. Wallace, and J. M. Anthony, *J. Appl. Phys.* **89**, 5243 (2001).
15. T. P. C. Juan, C. L. Lin, W. C. Shih, C. C. Yang, J. Y. M. Lee, D. C. Shye, and J. H. Lu, *J. Appl. Phys.* **105**, 061625 (2009).
16. S. W. Kang and S. W. Rhee, *Journal of The Electrochemical Society* **149** (6) C345-C348 (2002).
17. D. B. A. Rep and M. W. J. Prins, *J. Appl. Phys.* **85**, 7923 (1999).
18. C. H. Seager, D. C. McIntyre, W. L. Warren, and B. A. Tuttle, *Appl. Phys. Lett.* **68**, 2660 (1996).

19. K. Kodama, M. Takahashi, D. Ricinschi, A. I. Lerescu, M. Noda and M. Okuyama, *Jpn. J. Appl. Phys.* **41**, 2639 (2002).
20. T. P. Ma and J. P. Han, *IEEE Electron Device Lett.* **23**, 386 (2002).

Mater. Res. Soc. Symp. Proc. Vol. 1337 © 2011 Materials Research Society
DOI: 10.1557/opl.2011.980

New MEH-PPV Based Composite Materials for Rewritable Nonvolatile Polymer Memory Devices

Mikhail Dronov[1], Ivan Belogorohov[2] and Dmitry Khokhlov[3]
[1]A.M. Prokhorov General Physics Institute, Moscow, Russian Federation;
[2]Federal State Research and Design Institute of Rare Metal Industry ("Giredmet"),
Moscow, Russian Federation;
[3]M.V. Lomonosov Moscow State University, Moscow, Russian Federation.

ABSTRACT

We present the memory performance of devices with bistable electrical behavior based on MEH-PPV (Poly (1-methoxy-4-(2-ethylhexyloxy)-p-phenylenevinylene)) containing metal (Zn or Fe-Ni) particles. Another memory device based on aluminum phthalocyanine chloride (PcAlCl) added to the composite material reveals the photoinduced switching, in addition to the electrical one. Possible mechanisms for resistive switching are discussed.

INTRODUCTION

Resistive memory technologies currently receive much attention, being an advanced alternative to conventional memory devices. Organic materials are considered as a promising option for this sort of memory in view of a simple and low cost technology for their fabrication and an ability to produce flexible electronic devices. The resistive switching and the electrical memory effect in organic materials have been discovered in 1970s [1-2], but only recently, in the beginning of 2000s, have regained much attention [3-4]. Despite extensive experimental studies, no reliable theory explaining the switching effect was developed so far, nor any material or device with parameters attractive for a technology commercialization was presented [4].

We had chosen a MEH-PPV (Poly(1-methoxy-4-(2-ethylhexyloxy)-p-phenylenevinylene)) as a base element for a composite material because it is one of the most well studied conductive conjugated polymers. It is a commonly used constituent of OLEDs and photovoltaic devices. This material possesses well known and reproducible electrical properties. The idea to add metal particles into the functional material comes from the results of the paper [5] in which it was demonstrated that introduction of metal nanoparticles or a metal layer plays an important role in observation of conductivity switching in the composite material.

In our approach, a simple non-vacuum technology was used for formation of memory devices. Besides electrical switching, a possibility of light induced switching of conductivity was studied.

EXPERIMENT DETAILS

Device preparation

The main idea of our approach was to prepare samples using a simplest possible technology. Two types of samples were prepared. The first type was dip coated on an insulating substrate with already deposited contacts (fig.1a). The second type was prepared as bulk samples by casting into a form. In this case, two different contact configurations were used: the coplanar (fig.1b) and the sandwiched one (fig.1c). The distance between contacts varied in the range 100μm - 1mm, and the thickness of the samples was in the range 30 - 200μm. The contacts were made with silver paste. The composite materials were prepared by blending components (MEH-PPV, metallic particles, where two type of metallic particles – Zn or Fe-Ni alloy of 5μm mean size were used and for some series of samples the aluminum phthalocyanine chloride(PcAlCl) was added) in tetrahydrofuran (THF), and then treating the mixture in ultrasonic bath to insure the even distribution of components.

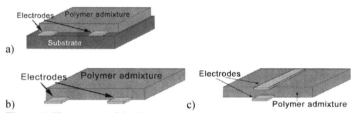

Figure 1. Three types of devices: a) on a substrate with coplanar electrodes; without substrate: b) coplanar electrodes, c) sandwiched electrodes

Measurements of the memory characteristics

An electrical analysis of the structures was performed using the Keithley 2612A Sourcemeter both in the pulse and constant current modes. The time resolved electrical measurements were performed using the Tektronix Signal Generator AFG3021B in the pulse mode and the Tektronix Digital Oscilloscope DPO 3054.

The role of sample geometry

It was found that devices with both sandwiched and coplanar contacts demonstrate almost the same conductivity, meaning that the silver paste contacts do not play any significant role in device properties, in particular, in the resistive switching. Therefore this effect is a feature of the material, not of the contact-material interface.

EXPERIMENTAL RESULTS

Electrically induced resistive switching

The initial state of the device is always insulating with the material conductivity estimated to be less than 10^{-5}S/cm in the electric field below 10V/cm. This value corresponds well to the conductivity of clean MEH-PPV samples. Application of higher voltage results in a unipolar switching. The threshold voltage varies only slightly within the samples series. It was found that a guaranteed switching to the conductive "ON" state is provided by application of the electric field ~ 500V/cm. The electric field of about 100-200V/cm is needed to switch the device back to the insulating "OFF" state. Application of voltage pulses exceeding the threshold ones leads to better R_{off}/R_{on} ratio, reaching as high as 10^7 (fig 2). The results are shown for a sample with Fe-Ni metal particles. The sample has coplanar electrodes, the sample thickness is 100µm, distance between the contacts is 200µm, the contact width is 900µm.

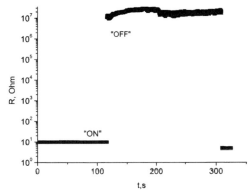

Figure 2. An example of resistive switching from the conductive to the insulating state and back.

A typical I-V curve of a device is presented in the fig.3. The sample is the same as in the Fig.2. The I-V curve is Ohmic in the "ON" state up to the current density of 1.5A/mm^2 without compromising the device performance and without switching back to the "OFF" state. The respective conductivity value reaches 10^2S/cm. At this high current density, some of the samples switched back to the insulating state or the devices burned out. However all samples stand successfully the current densities up to 0.5A/mm^2. A negative differential resistance region is clearly visible in the I-V curves at the current densities exceeding 1.5A/mm^2.

The characteristic switching times are small. Switching to the "ON" state occurs faster than for 100ns, and switching back to the "OFF" state is faster than 1µs.

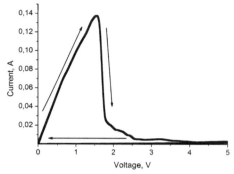

Figure 3. A typical I-V curve of a sample device

Retention time

Two types of retention time experiments were performed. In the first type of experiments, a write or erase pulse is followed by weak readout pulses or permanent readout with low voltage applied during up to 10 hours. More than 10^5 measuring pulses were done, demonstrating no change in the device state both in the conducting ("ON") and in the insulating ("OFF") mode. The second type of experiments implies periodic measurements of the device electric state during 3 months. This sort of experiments has demonstrated absence of any observable changes in the device electric state. Therefore it is reasonable to conclude that the memory devices under test demonstrated a reliable nonvolatile performance. Beside that, such a longtime nonvolatile behavior could be a manifestation of a filament formation as a possible memory mechanism, as discussed below.

Endurance experiments

Endurance testing was performed by repetitive application of write-read-erase-read cycles. The tests demonstrated the device stability for at least 10^4 cycles, as shown in the fig. 4. The results are shown for sample with Fe-Ni metal particles, coplanar electrodes, sample thickness 150μm, contact distance 300μm, contact width 1mm. This sample, however, has a lower particle concentration compared to the sample in the fig.2-3. Since we do not observe any noticeable degradation during these tests, it is possible to claim a reasonably good device performance. However further and longer studies are required to guarantee endurance of studied materials for RRAM operations.

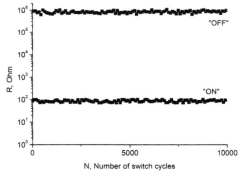

Figure 4. Endurance testing of MEH-PPV+(Fe-Ni) particles device.

Possible mechanisms of electrically controlled switching

One of the most probable switching mechanisms in our case is the filament formation [6-7], when switching is believed to occur due to formation of conductive filaments in the "ON" state. Switching to "OFF" state is due to the filament partial destruction. Some arguments in support of this mechanism provide the observed temperature dependencies of resistance in the "ON" and "OFF" states (fig.5, the results are shown for the same sample as in the fig.2-3) and the nonvolatile character of these states. The device demonstrates a metal-type temperature dependence of resistance in the "ON" state, whereas it is dielectric-like in the "OFF" state. On the other hand, no regions that could be interpreted as a result of Joule heating were observed in the I-V curves (see fig.3), so being most probable, the model of filaments formation is still questionable.

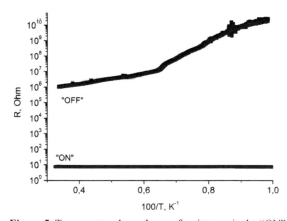

Figure 5. Temperature dependence of resistance in the "ON" and "OFF" states

Light controlled switching

A considerable photoresponse with R_{light}/R_{dark} up to 10^3 has been observed in the MEH-PPV+PcAlCl blends. The photoresponse relaxation time was on the order of several seconds at the room temperature.

The samples consisting of the same MEH-PPV+PhcAlCl blend with metal particles added a nonvolatile switching from the "OFF" to the "ON" state; this switching was observed at maximal radiation intensity. A lower level of illumination provided a strong decrease in the switching voltage. The values of V_{write} and V_{erase} decreased by more than 10 times, so as low as few hundreds of mV applied voltages were enough for a device to switch its state.

Mechanisms for the light induced switching are unclear. However the results with lower radiation levels show that threshold voltages may be significantly reduced. There is a possibility that spatial non-homogeneities in irradiation with light may lead to a non-uniform generation of extra charges resulting in appearance of a photovoltaic effect providing, in turn, switching of a sample state.

SUMMARY

A new material with outstanding characteristics is proposed for application as a material for RRAM devices. In this material, the switching ratio of up to 10^7 was achieved, with the minimal ratio of 10^5 in the normal operation mode. The current density in the "ON" state was demonstrated to reach $1.5 A/mm^2$ in some samples, and about $0.5 A/mm^2$ on a regular basis. Long retention rates of more than 3 months and good reliability for more than 10^4 rewrite cycles have been demonstrated. A good nonvolatile behavior is expected if the material properties are retained while scaled down to smaller dimensions.

The effect of light induced resistive switching was observed. This effect opens a new direction in RRAM investigations, it can be used for the opto-electric circuiting in the modern and future electronics.

REFERENCES

1. H. Carchano, R. Lacoste, Y. Segui, Appl. Phys. Lett. 1971, **19**, 414.
2. L. F. Pender, R. J. Fleming, J. Appl. Phys. 1975, **46**, 3426.
3. D. Prime, S. Paul, Phil. Trans. R. Soc. A 2009 **367**,4141-4157.
4. J. C. Scott, L. D. Bozano, Adv. Matter. 2007, **19**,1452-1463.
5. L. D. Bozano, B. W. Kean, M. Beinhoff, K. R. Carter, J. C. Scott, Adv. Funct. Mater. 2005, **15**, 1933.
6. G. Dearnaley, D. V. Morgan, A. M. Stoneham, J. Non-Cryst. Solids 1970, **4**, 593.
7. G. Dearnaley, A. M. Stoneham, D. V. Morgan, Rep. Prog. Phys. 1970, **33**, 1129.

Mater. Res. Soc. Symp. Proc. Vol. 1337 © 2011 Materials Research Society
DOI: 10.1557/opl.2011.1126

Planar Non-Volatile Memory based on Metal Nanoparticles

A. Kiazadeh[1], H. L. Gomes[1], A. R. Da Costa[2], P. Rocha[2], Q. Chen[2], J. A. Moreira[2], D. M. De Leeuw[3] and S. C. J. Meskers [4]

[1] Center of Electronics Optoelectronics and Telecommunications (CEOT)
[2] Centro de Investigação em Química do Algarve
Universidade do Algarve, Campus de Gambelas, 8000-139 Faro, Portugal,
[3] Philips Research Laboratories, High Tech Campus 4 WAG 11, 5656 AE Eindhoven, The Netherlands
[4] Molecular Materials and Nanosystems, Eindhoven University of Technology, P.O. Box 513, 5600 MB Eindhoven, The Netherlands

ABSTRACT

Resistive switching properties of silver nanoparticles hosted in an insulating polymer matrix (poly(N-vinyl-2-pyrrolidone) are reported. Planar devices structures using interdigitated gold electrodes were fabricated. These devices have on/off resistance ratio as high as 10^3, retention times reaching to months and good endurance cycles. Temperature-dependent measurements show that the charge transport is weakly thermal activated (73 meV) for both states suggesting that nanoparticles will not aggregate into a metallic filament.

INTRODUCTION

Metal-Insulator-Metal (MIM) structures where the insulator layer is comprised of metal nanoparticles (Nps) or nanostructured metal films embedded in semiconductive or semi-insulating host matrices have attracted considerable attention due to their interesting electrical properties and because of their potential for high density non-volatile memory applications. These devices show dramatic changes of the electrical resistance, so-called resistive switching effect. Current-voltage characteristics switch reversibly between a low conductance off-state and a high conductance on-state [1-7]. This phenomenon has being intensively investigated worldwide for developing Resistive Random Access Memories (RRAMs), a possible competitor, and even replacement for flash memory and hard-disk drives. It is relatively easy to produce colloidal solutions a well-defined contribution of nanoparticles and polymers, and it is the advantage of nanoparticles based systems. Thin films can be deposited by spin-coating, printing, or dip-coating techniques offering the prospect of low fabrication cost, mechanical flexibility and light weight.

In spite of intense efforts, the detailed physics of the resistive switching is still not elucidated. The majority of the work reported in literature, is for sandwich type structures where one of the electrodes is often a reactive metal such as aluminum [8]. This makes difficult to discriminate interfacial effects occurring at the electrode/material interface from bulk process. In this contribution, we fabricate a planar structure using interdigitated gold electrodes. The resistive switching channel is comprised of silver nanoparticles embedded into an insulating polymer matrix. This planar structure is much simpler than sandwich structures and may contribute to elucidate the detailed physics of the resistive switching phenomena. Furthermore, planar structures have a significant lower intrinsic capacitance than sandwich structures, therefore, they should have a faster dynamic response. In addition, planar structures have the active layer directly exposed and can be probed by a number of surface analytical techniques to identify and characterize topographical, morphological changes that occur in the devices upon resistive switching.

EXPERIMENT

Poly (N-vinyl-2-pyrrolidone) (PVP, MW=25000) was supplied by Fluka. Other chemicals were purchased from Merck and used without any further purification. Ultraviolet absorption spectrum measurements were carried out with an (Shimadzu-UV 2550-8030) Spectrophotometer. Transmission Electron Microscopy (TEM) was performed with a CM-200 FEG Philips apparatus.

Silver nanoparticles were prepared by reducing the silver nitrate in poly-vinyl-pyrrolidone (PVP) aqueous solution. Dimethylformamid DMF was used as a reducing agent. The PVP solution was prepared by dissolving PVP in DMF and stirred at 300 rpm then $AgNO_3$ solution with considering 1:4 molar ratio of silver to PVP was added into the solution. The formation of silver nanoparticles was confirmed by UV-Vis spectrophotometer. The average size (25 nm) was estimated by transmission electron micrograph TEM. Figure 1 show the device test structure used to measure the electrical properties of the system comprised of nanocrystals in a semi insulating or insulating host matrix. A thin film is formed between the two gold electrodes on top of the insulating silicon dioxide surface by spin coating. Prior to electrical measurements the samples were also pumped in vacuum to remove further any residual solvent. Electrical measurements were carried out using a Keithley 487 picoammeter/voltage source in dark and in high vacuum conditions. To test the electrical properties of these structures we applied symmetrical voltage ramps with increasing voltage. During all the measurement the conductive silicon substrate is kept grounded to prevent charging of the SiO_2 layer.

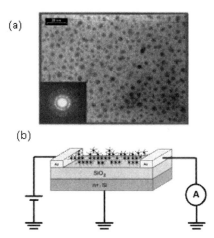

Figure 1 (a) TEM image of PVP capped silver nanoparticles (b) and planar device structure with 10 μm separated gold electrodes.

DISCUSSION

Initially, the system comprised of silver nanoparticles embedded in the PVP matrix behaves as an insulating or semi-insulating material. The corresponding current-voltage-characteristic is the lower curve represented in figure 2. The resistance at 10 V is approximately 10 GΩ. This insulating state of the sample is named pristine state. Then suddenly, near 45 V the current rises dramatically and switch to the upper *I-V* characteristic represented in figure 2. This state is named electroformed state.

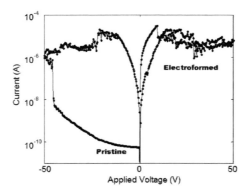

Figure 2 Current-voltage characteristics for the Pristine and the electroformed states.

The change in device electrical properties from pristine state to more conductive state is irreversible if the device is kept under vacuum. This permanent modification corresponds to an electroforming process [9]. The nature of the electroforming process in nanoparticles-based systems is still not clear. After subsequently *I-V* loops the noisy *I-V* characteristic observed immediately after the switch from a pristine to an electroformed state (see figure 2), evolves to a noise free on-state characteristic with well-defined negative differentially resistance (NDR) regions located at around |35| V as represented figure 3. The local maximum and local minimum of its current is indicated as Von and Voff, respectively. The programming of the two conductive states is done by a voltage pulse with a length of a few seconds. The on and off-state can be set by voltages pulses of either polarity. A pulse near Voff will bring the device to the low conductance state or off-state. The high conductance curve (on-state) is restored by using a pulse near Von which is typical 30 V. The on/off ratio is almost 3 orders of magnitude as shown in figure 3. The memory is non-volatile. The retention time of both states was measured for a period of 10 days as shown in figure 4(a). The on/off ratio does not degrade with time. The cycle endurance represented in figure 4(b) was measured by programming successively the memory. The voltage pulse sequence used was as follow: write voltage (W) is 30 V and the erase (E) is 50 V. For both states the read voltage (R) is 3 V. The temperature dependence of the on and off states was measured in the temperature range of [320-200 K] with a continuous applied voltage of 3V after pulses near to Voff and Von for each state. The Arrhenius plot in figure 5 shows that the current is weakly thermally activate with single activation energy for both states of only ~73 meV.

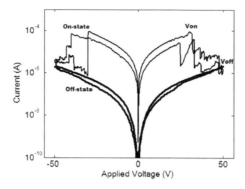

Figure 3 Current-voltage characteristics for the on and off states. The on-state shows negative differential regions (NDR).

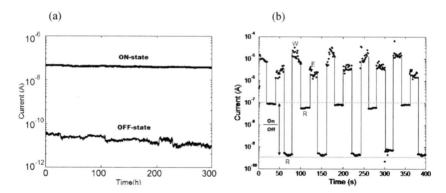

Figure 4 (a) Retention time for both states measured at 3V (b) Cycles of endurance to show the programmable device at both on and off states. The write voltage (W) is 30V and the erase voltage (E) is 50V, for both states the read voltage (R) is 3V.

Although, it is low activation energy, it still suggests that metallic paths are not established between the electrodes; otherwise we would expect a non-activated behavior or even a positive temperature coefficient for the change in current. Switching has been reported in sandwich structures of AgI by Liag *et al.* [10] and for Ag2S by Waser and Aono [11]. Both reports treated the switching layer as a solid electrolyte where silver ion migration is claimed to occur with an electroactive electrode. Resistive switching is then readily explained by formation and rupture of metallic filaments. The findings in planar structures reported here seems to discharge this process, because the electrodes are considerable apart (10 μm).

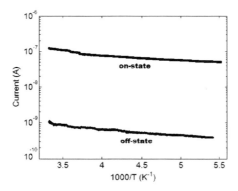

Figure 5 Temperature dependence of the current for the on and off states measured at 3V.

Furthermore, they are non-reactive gold electrodes; migration of metallic species from electrodes is unlikely to occur. It is interesting to point out that the switching behavior of these planar memory devices is similar to the switching behavior of sandwich type metal-oxide polymer diodes [12]. Both systems need an electroforming process; they show almost undistinguishable electrical properties, similar current-voltage characteristics with NDR regions, and equal programming features. This suggests that the underlying resistive switching mechanism is also similar. It is possible that the silver nanoparticles may form silver oxide clusters distributed within the polymer matrix. Therefore, the two switching systems become also physically similar.

CONCLUSIONS

The electrical switching and memory effects of the device made from AgNPs/PVP film are investigated. The device can be switched between high and low conductance states by voltage pulses. The endurance characteristics were obtained for ~ 1000 programmed cycles with write and erase voltages of 30 V and 50 V, respectively. The retention time measurements show the memory states are stable for months when kept under vacuum. The on and off-states are stable for months as long as the device is kept under vacuum. Under ambient atmosphere it shows different switching hysteresis (unpublished data). The respective conductance states are volatile and unreliable, suggesting that they are related with surface electrical conduction mediated by moisture. The temperature measurements show that for both on and off states the current is weakly thermal activated, suggesting that metallic conduction does not occur in this structure. These devices are free of complex electrode interfaces, they have the conduction channel accessible to surface analytical techniques, therefore, they are an ideal tool to get insight into the fundamental aspects related how metallic nanoparticles contribute for the charge carrier trapping and transport in a insulating matrix.

ACKNOWLEDGMENTS

We gratefully acknowledge the financial support received from the Dutch Polymer Institute (DPI), project n.º 703, from Fundação para Ciência e Tecnologia (FCT) through the research Unit, Center of Electronics Optoelectronics and Telecommunications (CEOT), REEQ/601/EEI/2005 and the POCI 2010, FEDER

REFERENCES

1. L. D. Bozano, B. W. Kean, M. Beinhoff, K. R. Carter, P. M. Rice, J. C. Scott, J. Adv. Fun. Mater. 15, 1933 (2005)

2. G. Liu, Q. D. Ling, E. T. Kang, K. G. Neoh, D. J. Liaw, F. C. Chang, C. X. Zhu, D. S. H. Chan, J. Appl. Phys. 102, 024502 (2007)

3. S. Paul, A. Kanwal, M. Chhowalla, J. Nanotechnology, 17, 145 (2006)

4. H. S. Majumdar, J. K. Baral, R. Österbacka, O. Ikkala, H. Stubb, J. Org. Electron. 6, 188 (2005)

5. H.T. Lin, Z. Pei, Y. J. Chan, IEEE Electron Dev. Lett. 28, 569 (2007)

6. B. Pradhan, S. K. Batabyal, A. J. Pal, J. Phys. Chem. B. 110, 8274 (2006)

7. B. Cho, T. W. Kim, M. Choe, G. Wang, S. Song, R. Lee, J. Org. Electron. 10, 473 (2009)

8. J. C. Scott, L. D. Bozano, J. Adv. Mater. 19, 1452 (2007)

9. J. G. Simmons and R. R. Verderber, Proceedings of the Royal Society of London, Series A. 301, 77 (1967)

10. X. F. Liang, Y. Chen, L. Shi, J. Lin, J. Yin and Z. G. Liu, J. Phys. D: Appl. Phys. 40, 4767 (2007)

11. R. Waser and M. Aono, J. NatureMaterials. 6, 833 (2007)

12. F. Verbakel, S. C. J. Meskers, R. A. J. Janssen, H. L. Gomes, M. Cölle, M. Büchel, D. M. de Leeuw, J. Appl. Phys. Lett. 91, 192103 (2007)

AUTHOR INDEX

SUBJECT INDEX